PROFESSOR R. W. FOGEL
HARVARD UNIVERSITY
1737 CAMBRIDGE ST., RM. G-7
CAMBRIDGE, MA. 02138

12/16/80

World Food Supply

This is a volume in the Arno Press collection

World Food Supply

Advisory Editor
D. Gale Johnson

Editorial Board
Charles M. Hardin
Kenneth H. Parsons

See last pages of this volume for a complete list of titles.

HIGH-YIELDING
VARIETIES OF GRAIN

ARNO PRESS
A New York Times Company
New York — 1976

Editorial Supervision: MARIE STARECK

Reprint Edition 1976 by Arno Press Inc.

Copyright © 1976 by Arno Press Inc.

WORLD FOOD SUPPLY
ISBN for complete set: 0-405-07766-1
See last pages of this volume for titles.

Manufactured in the United States of America

Library of Congress Cataloging in Publication Data
Main entry under title:

High-yielding varieties of grain.

(World food supply)
Reprint of The impact of new grain varieties in Asia, by J. W. Willett, first published 1969, by U. S. Dept. of Agriculture, Economic Research Service, Foreign Regional Analysis Division, Washington; of High-yielding varieties of wheat in developing countries, by S. K. Tsu, first published 1971, by Dept. of Agriculture, Economic Research Service, Washington; of Accelerating India's food grain production, 1967-68 to 1970-71: requirements and prospects for a yearly growth rate of 5 percent, by W. E. Hendrix, J. J. Naive, and W. E. Adams, first published 1968, by U. S. Dept. of Agriculture, Economic Research Service, Washington; and of Technological change in agriculture: effects and implications for the developing nations, by D. G. Dalrymple, first published 1969, by Foreign Agricultural Service, U. S. Dept. of Agriculture, Washington.
  Includes bibliographies.
  1. Grain--Varieties--Addresses, essays, lectures.
2. Agricultural innovations--Addresses, essays, lectures.
3. Underdeveloped areas--Agriculture--Addresses, essays, lectures. 4. Grain--Asia--Addresses, essays, lectures.
5. Grain--India--Addresses, essays, lectures.
  I. Series.
SB189.H53   633'.1'047   75-26315
ISBN 0-405-07800-5

CONTENTS

Willett, Joseph W.
THE IMPACT OF NEW GRAIN VARIETIES IN ASIA (Economic Research Service-Foreign 275, United States Department of Agriculture), Washington, D. C., 1969

Tsu, Sheldon K.
HIGH-YIELDING VARIETIES OF WHEAT IN DEVELOPING COUNTRIES (Economic Research Service-Foreign 322, United States Department of Agriculture), Washington, D. C., 1971

Hendrix, William E., James J. Naive and Warren E. Adams
ACCELERATING INDIA'S FOOD GRAIN PRODUCTION, 1967-68 TO 1970-71: Requirements and Prospects for a Yearly Growth Rate of 5 Percent, (Foreign Agricultural Economic Report No. 40, Economic Research Service, United States Department of Agriculture), Washington, D. C., 1968

Dalrymple, Dana G.
TECHNOLOGICAL CHANGE IN AGRICULTURE: Effects and Implications for the Developing Nations (Foreign Agricultural Service, United States Department of Agriculture in cooperation with Agency for International Development), Washington, D. C., 1969

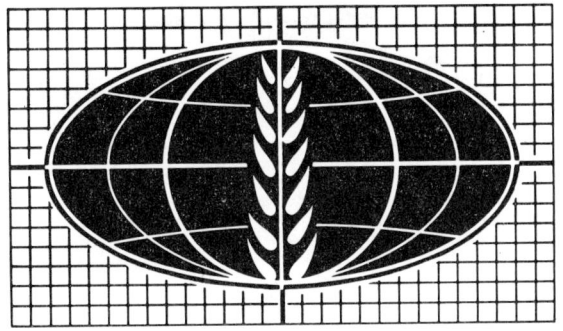

THE IMPACT OF NEW GRAIN VARIETIES IN ASIA

ERS-FOREIGN 275

U.S. DEPARTMENT OF AGRICULTURE
ECONOMIC RESEARCH SERVICE
FOREIGN REGIONAL ANALYSIS DIVISION

PREFACE

This report on the impact of new varieties of grain in Asia was prepared by the Economic Research Service under an agreement with the Agency for International Development. It is part of a larger study on the outlook for demand of agricultural products produced by the less developed countries. An earlier version was presented at The Spring Review conference of the Agency for International Development, Washington, D.C., May 13-15, 1969. This report benefited from the many excellent studies prepared for that conference, as well as from discussions by the participants; but the analysis and conclusions contained herein are entirely the responsibility of the author.

This report estimates the contribution of the new varieties of rice and wheat to the production of grain in Asia in 1968-69. Since the estimate is necessarily rough, it is presented as a broad range of possibilities. The paper includes a discussion of various factors which may tend to slow the spread of the new varieties in Asia and also comments on some of the probable economic and social effects associated with these new varieties.

Special thanks are due Donald Chrisler and Dana G. Dalrymple who rendered invaluable service in many ways in preparing the report.

CONTENTS

	Page
Summary	5
The Setting	7
How the New Varieties Were Developed	8
The Mexican wheat varieties	8
The IRRI rice varieties	9
Some Characteristics of the New Varieties	9
Area Planted to the New Varieties	10
Yields and Production Increases	12
Yields	12
Production increases	13
Multiple Cropping	15
Limitations to Spread in Individual Countries	16
Irrigation	16
Risks of diseases and pests	18
Prices and incentives	18
The Beneficiaries—the People Left Behind	19
Technological innovations	19
The people left behind	20
The Outlook for World Supply and Demand	21
Demand—self-sufficiency	21
Shortrun outlook for grain exports of the less developed countries	22
The longer run outlook for international grain markets	23
Bibliography	24

Washington, D.C. 20250   July 1969

SUMMARY

In the last 2 years, the rapid spread of highly productive new varieties of rice and wheat in several less developed countries of Asia has increased the likelihood that these countries will improve the diets of their rapidly growing populations. These new varieties of grain, along with better weather, more fertilizer, higher prices to farmers, and other factors, have helped to bring about dramatic increases in grain production in India, Pakistan, and the Philippines.

The new wheat was developed in the late 1950's in Mexico. The rice varieties were developed in the early 1960's at the International Rice Research Institute in the Philippines. Both types of grains have short, stiff straw and produce much higher yields than traditional varieties without lodging. Their adaptability to wide differences in latitude has contributed to their rapid spread.

The new grains can make better use of larger amounts of fertilizer than traditional varieties, but their water requirements are high. The shortage of irrigation systems with adequate water supplies may be the most critical physical factor limiting the further spread of the new grain varieties.

The new seeds produce crops in shorter time periods than most traditional varieties, and thus sometimes make it possible to raise additional crops.

The new varieties have several disadvantages. Wet grain has been a problem, necessitating investment in artificial drying equipment in some instances. Moreover, since consumers rate the new grains rather low on palatability, they have been marketed at considerable discounts. Because of their rapid introduction, these grains may yet prove to be susceptible to native pests and diseases.

The rapid spread of the new varieties has generally been due to vigorous government programs. U.S. aid programs have assisted in a variety of ways. The swift adoption of these grains has clearly demonstrated that farmers in the less developed countries will readily accept new practices when the inputs are available and returns are substantial.

The improved wheat is used on almost all of Mexico's wheat areas. Both the new wheat and new rice varieties have spread rapidly in Asia. In 1968/69, they occupied about 7 percent of the riceland and about 16 percent of the wheatland in the less developed areas of Asia (excluding Communist China). It is estimated that, under average weather conditions, they would add about 9 percent to rice production and 20 percent to wheat production in the area, based on the judgment that their yields are from 30 to 100 percent greater than traditional varieties raised under similar irrigated conditions.

The future spread and production of the new varieties is uncertain. Factors which will influence their adoption include prices of grain and inputs, extension and improvement of irrigation systems, and damage from pests.

Farmers in a position to adopt the new varieties quickly can benefit substantially. Others may be harmed by the increased competition. The effects on farm labor are uncertain. In some situations the new grains may increase demand and thus (at least temporarily) bring about a rise in the wages of farm labor. On the other hand, the corresponding stimulus to mechanization may displace some labor.

The evaluation of the effects of the increased production on consumers requires judgments as to whether the increases will add to net supplies, replace food aid, displace commercial imports, or perhaps, to some extent, be exported. The actual situations will depend upon government decisions as well as market forces.

It seems possible that the less developed countries of Asia can become nearly self-sufficient in grain before many years. However, if supplies continue to grow faster than demand after self-sufficiency is attained, prices will be depressed, unless international markets can absorb additional increases. World supplies of grain have been growing relative to demand. The high producer prices which have been incentives to the rapid adoption of the new technology probably will be impossible to maintain. The poorer countries generally cannot afford to subsidize the production of grain, either to sell domestically at low prices or to export. Already some surpluses have arisen.

The longer run outlook for grain exporters does not appear promising. Surpluses of grain, or the capacity to produce surpluses, especially wheat, seem likely to grow. Government policies in both the developed and less developed countries, with respect to domestic agricultural programs, trade, and aid, will be vital in determining the economic framework in which further increases in production of grain will take place.

THE IMPACT OF NEW GRAIN VARIETIES IN ASIA

by

Joseph W. Willett 1/

THE SETTING

Less than 2 years ago, concern was expressed by a number of commentators that the world was losing the race between population and food production. The evidence usually cited was a comparison between "recent trends" in population and food production, shifts in patterns of grain trade, and a decline in surplus grain stocks, especially those held by the United States. There were even predictions of impending mass starvation.

These pessimistic views about the longrun outlook for food and agriculture were reinforced by shortrun developments. By 1966, world grain stocks were rapidly drawn down, mainly as a result of expanded imports by India and the Soviet Union. India had suffered two droughts in succession, and the Soviet Union had two crop failures in 3 years. Australia also had a poor crop. U.S. stocks were greatly reduced, the outlook for wheat yields was unfavorable, and the acreage allotment for the 1967 U.S. wheat crop was increased considerably.

As a result, the United States made considerable changes in its aid policy. In addition to lending support for population control, more emphasis was placed on agricultural development. Countries receiving assistance were strongly encouraged to place a higher priority on agricultural production.

This crisis atmosphere was an important factor in the rapid adoption of new varieties of grain in Asia. The governments of some Asian countries became especially concerned over their food problems, and the availability of highly productive grains seemed to offer a solution. In some instances, governments mounted high-priority programs to ensure that farmers would quickly adopt the new seed. These programs helped to make the essential inputs available to farmers to grow the new grains. The availability of the inputs together with the incentive of high grain prices quickly spread the new seed over substantial acreages.

Apocalyptic predictions of food shortages are still being made, but in the past year and a half opinions have been expressed that the world food outlook has changed radically. Evidence cited to support this view includes falling grain prices, increased grain stocks, and the rapid spread of the new grains.

1/ Deputy Director, Foreign Regional Analysis Division, Economic Research Service.

Dramatic increases in production have occurred in India, Pakistan, and the Philippines. In these and several other of the less developed countries of Asia, new varieties of rice and wheat have become significant in a very short time.

For several reasons, it is not possible to estimate accurately either the actual impact of the new grains thus far or their potential. In most less developed countries, statistics on acreage, production, and yields of crops, including grain, are not reliable. The term ''new varieties'' is used to include varying groups of grains. The new grain varieties have not yet been widely tested in most countries, and little information on farm experience is available. In many of the less developed countries, the weather causes wide year-to-year variations in yield. These variations complicate the analytical problem of isolating the effects of the new varieties. Exceptionally high grain prices in recent years have also been influential in the generally enthusiastic reception given the new varieties.

HOW THE NEW VARIETIES WERE DEVELOPED

The new grain varieties which have received the most publicity in the last few years, and which are already making a substantial contribution to increased production in Asia, are the so-called ''Mexican'' wheats, and rice varieties developed and disseminated by the International Rice Research Institute (IRRI), especially those designated as IR-5 and IR-8. These grains were developed in a remarkably short time. However, the breeders of the new varieties had much firmly established earlier work on which to draw.

The Mexican Wheat Varieties

In the early 1940's, a program was initiated in Mexico which concentrated first on improving varieties and increasing production of corn and wheat and was later expanded to cover other commodities. Rust was a major factor limiting the yields of wheat in Mexico, and early breeding efforts were directed to developing resistant varieties (4, p. 3, 26, p. 239). 2/ Later, the breeders concentrated on producing wheat with short, stiff straw and a high response to fertilizer (26, p. 239).

Norin 10, a variety of wheat with short, stiff straw, which was first registered in 1935, was developed by the Japanese by crossing two U.S. varieties with several Japanese varieties. S. C. Salmon, who worked with the Japanese after World War II, brought Norin 10 to the United States where it was distributed to wheat breeders in 1947-48. Orville A. Vogel, a USDA scientist working in the State of Washington, used Norin 10 in developing Gaines wheat, a variety which has set world yield records in the northwest region of the United States. In 1953, Norman E. Bourlaug, a Rockefeller scientist working in Mexico, obtained some wheat varieties with short straw from Vogel. Norin 10 made a major contribution of germ plasm to the new Mexican wheats (26, pp. 236-238).

2/ Underscored numbers in parentheses refer to items in the Bibliography, page 24.

The IRRI Rice Varieties

The new varieties of rice are Indicas. The Indicas have long been predominant in tropical Asia, although their yields have been much lower than the Japonicas of Japan and Taiwan. Despite their higher yields, several characteristics of the Japonicas have prevented their spread into the tropical areas. They are more prone to disease than the native Indicas; they do not thresh well using the methods commonly employed in the Asian tropics; consumers generally do not care for their taste; and they lack a period of seed dormancy. The latter is especially important to prevent germination in areas where the harvest takes place during a rainy period (4, p. 17).

The International Rice Research Institute, which is supported by the Ford and Rockefeller Foundations, was dedicated in 1962 at Los Banos in the Philippines. During 1962, IRRI rice breeders made a number of crosses involving tall, tropical Indica varieties, the Ponlai Japonica variety from Taiwan, and several semidwarf Indica varieties from Taiwan. By 1965, IR-8 had been developed and given its first yield trial. IR-8 was obtained by crossing Peta, a tall rice from Indonesia, with Dee-geo-woo-gen, a short rice from China (8, pp. 252-253).

SOME CHARACTERISTICS OF THE NEW VARIETIES

Because of the genetic characteristics which have been bred into the new varieties of rice and wheat, these grains produce much higher yields than traditional varieties when conditions are favorable. The short, stiff stems are important in achieving increased productivity under heavy fertilization, because the plants do not lodge, or fall over, when heavy applications of fertilizer produce a heavy seed head. Grain which has lodged does not develop properly and is harder to harvest; photosynthesis is interfered with, yields are reduced, and growth of molds is stimulated. Generally, height is reduced when the growing period is shortened (4, p. 16). The new varieties accomplish far more photosynthesis then traditional varieties during the period when the grain is produced. Also, the ratio of grain to straw is greatly increased in comparison with older varieties.

In developing IR-8, thus far the most important of the new rice varieties, breeders reduced the height to 100 centimeters, compared with a height of perhaps 180 centimeters for traditional varieties. The short, upright leaves of IR-8 permit water to run off quickly and allow sunlight to penetrate to the lower leaves. The straw of the plant is not only reduced in length, but it is also exceptionally stiff because the breeders selected plants which had thick stems wrapped with leaf sheaves (8, pp. 254-255). Under some circumstances, the short, stiff stems of the new varieties may be a disadvantage. In East Pakistan, for example, plantings are made on land which is subject to uncontrolled flooding; the native varieties are better able to withstand such conditions.

The Mexican wheats and IR-8 have proved to be productive in areas with wide variations in the length of day. This adaptability has been important in their rapid spread to different latitudes (26, p. 239). The breeders have also incorporated into the new rice varieties the ability to produce many stems on a single plant, which lowers seed requirements (8, p. 252).

The new varieties of rice ripen in 120-125 days, rather than the 180 days required by most traditional varieties (30, p. 2). The shorter growing season increases the possibilities of producing more crops per year on the same land. However, a shorter growing period also sometimes brings the grains to maturity during the wet season. Thus, the customary method of spreading rice on the ground to dry may be inadequate and artificial drying may be required (6, p. 693).

The new rice varieties have some characteristics which make them relatively undesirable for processors and consumers. They do not mill as satisfactorily as older varieties, and consumers generally rate them lower on palatability. Although these disadvantages probably can be bred out, both the new rice and wheat varieties have often sold considerably below the prices of traditional grains (10, p. 44).

AREA PLANTED TO THE NEW VARIETIES

In some areas, the new grain varieties have spread with extraordinary rapidity. This rapid dissemination has been built on a well-established institutional base (as in India's Intensive Agricultural Districts Program), vigorous government action (as in Turkey), or both. Although the existence of an institutional base seems to have been important, many ad hoc arrangements have been employed. In some cases, institutions were used to perform jobs other than their traditional ones (27, p. 28). The availability of a combination or "package" of inputs at subsidized prices has greatly stimulated farmer acceptance of the programs. The rapid spread has clearly demonstrated that farmers in less developed countries will quickly and enthusiastically adopt new methods if the inputs are available and the benefits are substantial. AID's role in the programs has been substantial, but has varied according to circumstances in the individual countries.

The improved wheat varieties spread rapidly on the irrigated acreage of the wheat farmers of Mexico from 1949 to 1956. At present, the improved seed is used on nearly all of Mexico's wheat area, of which nearly 90 percent is irrigated and more than two-thirds is fertilized (11, p. 14).

The Mexican wheats were introduced into Pakistan and India in small quantities during 1963/64 and tests were conducted. In 1966, India made a large purchase of seed from Mexico for planting in the fall of that year. In 1967, Pakistan made an even larger purchase (7, p. 90).

Despite their rapid spread, the new varieties were not planted on a large enough share of the grain acreage in Asia in 1967/68 to have had a major impact on production in the less developed countries as a whole. They did affect production in certain regions and in certain countries, however. As indicated in table 1, less than 3 percent of the rice area in South and Southeast Asia was planted to new varieties in 1967/68. Wheat occupies a far smaller share of crop acreage than does rice in the less developed Asian countries, but about 11 percent of the wheat area in West and South Asia was planted to new varieties in 1967/68. The country-to-country variation in the rate of adoption is great. The share of the total wheat area seeded to new varieties in 1967/68 was insignificant in Turkey and Afghanistan but amounted to about 12 percent in Pakistan and nearly 20 percent in India.

Table 1.--Estimated area planted to new varieties of rice and wheat in West, South, and Southeast Asia, 1966/67-1968/69 1/

Country or region	Rice 1966/67	Rice 1967/68	Rice 1968/69 2/	Wheat 1966/67	Wheat 1967/68	Wheat 1968/69 2/
	---------- Million hectares ----------					
Turkey					2/0.17 (8.1)	0.60
Iran					(4.2)	
Afghanistan					.02 (2.3)	
Nepal				0.01	.02 (.1)	
West Pakistan		3/(1.4)	0.28	.11	.73 (6.0)	1.21
East Pakistan		0.06 (9.9)	.08	2/ .01 (.1)		.02
India	0.87	1.78 (36.7)	3.77	.52	2.94 (14.9)	4/4.05
Burma		(5.2)	.22			
Thailand		(6.1)				
South Vietnam		(2.3)	.04			
Philippines	.07	.24 (3.0)	.45			
Indonesia		(7.4)	.38			
Total	.94	2.08 (72.0)	5.22	.64	3.89 (35.7)	5.88
Other countries		(5.3)			(.6)	
Total rice area, South and Southeast Asia		(77.3)				
Total wheat area, West and South Asia					(36.3)	

1/ Adapted from Dalrymple, Dana G., **Imports and Plantings of High-Yielding Varieties of Wheat and Rice in the Less Developed Nations** (unpublished), U.S. Dept. Agr., IADS, Dec. 17, 1968.
2/ Target or projection.
3/ Figures in parentheses are total area.
4/ Given as 2.63 in India Program Memorandum, FY 1970, US/AID Mission, New Delhi, Sept. 1968, p. D-55.

In the 1968/69 crop season, about 5 million hectares were planted to the new rice varieties in South and Southeast Asia, or nearly 7 percent of the total riceland in the region. The new wheat varieties were scheduled for planting on 6 million hectares, or about 16 percent of the total wheat area in West and South Asia.

YIELDS AND PRODUCTION INCREASES

Yields

Reliable estimates of the impact of the new varieties on grain availability in Asia would require the comparison of yields of new and traditional varieties by region, while accounting for the influence of such factors as weather, acreage changes, fertilizer, prices, and availability of credit, irrigation, and extension services. The new varieties are generally produced under the best possible conditions. Therefore, it would be misleading to compare their yields with yields of traditional varieties grown under a variety of conditions and then assign all the differences to the genetic characteristics.

Following are some aspects of the problem of evaluating yields of the new grains:

(1) The new varieties were first planted on substantial acreages in Asia in 1967/68, a year of generally good weather in the region. In India and Pakistan, the large increases in area planted to new varieties in that year coincided with excellent weather, following 2 years of drought. It has been estimated that good weather contributed more than did the new varieties to India's increase in grain production in 1967/68 (33, p. D-20). During the 1967/68 wheat growing season in West Pakistan, average monthly rainfall was 135 percent above that of the previous two seasons (14). Turkey also experienced good weather in 1967/68 and correspondingly high yields.

(2) In 1967/68, the total acreages planted to both new and traditional varieties increased as follows: wheat in India, 10 percent; wheat in Pakistan, almost 10 percent (irrigated wheat acreage expanded 20 percent) (14); and rice in Pakistan, almost 10 percent. Rice acreage remained unchanged in India and, perhaps, in the Philippines. 3/ Turkey's wheat area did not change.

(3) There have been frequent references to planting the best land to new varieties. In Turkey, Mexican wheats were distributed in ''areas of low altitude and high rainfall'' (35, p. 5). In the Philippines, IR-8 was planted on the ''most productive rice growing area'' (28, p. 4). All of the new wheat varieties in Pakistan were planted on irrigated land. This apparently was also the case in India where the Mexican wheats are reported to have been cultivated by better-than-average farmers.

(4) Fertilizer consumption increased 50 percent in India and 30 percent in Pakistan in 1967/68. It is reasonable to assume that much of that increment was used on new varieties, but detailed information is not available.

3/ Estimates of 1967/68 Philippine rice acreage vary by 10 percent.

With adequate water, controlled irrigation, and other improved practices, the increased response of the new varieties to fertilizer can be substantial. The actual yield advantage will depend upon the level of fertilization, which, in turn, will depend upon the farmers' incentives to use fertilizer. The latter will be a function of the price of fertilizer and the price of grain. The yield advantages of the new varieties are generally greater at high levels of fertilizer use; when no fertilizer is used, they seem to have little if any advantage over traditional varieties. Thus, the yield advantages of the new varieties will be relevant only when the economic situation provides an incentive to use fertilizer. For example, one analysis has concluded that with the traditional wheat varieties used in India in 1963, fertilizer (with an ''optimum'' application of 58 kilograms per hectare) holds a promise of only a 50-percent yield increase. But Sonora Wheat 63 can use more fertilizer effectively and (with the application of 116 kilograms per hectare) will give a doubling of yields. Thus, it may be said that Sonora 63 has a 30-35 percent yield advantage over the traditional 1963 varieties (200 percent divided by 150 percent); however, Sonora required twice the dosage of fertilizer applied to the traditional varieties to achieve this yield (32, p. 695). Although there have been a few farm management studies of costs and returns, there is insufficient basis to generalize to aggregate supply functions which would predict the overall response of output to price changes (10, p. 45).

Yield experiences with the new grains have varied greatly, according to farmers' reports (18, p. 7). Numerous experiments and one farmer survey suggest that the new rice and wheat varieties have a yield advantage of 30-100 percent, when planted with adequate irrigation and a high level of fertilization and compared with traditional varieties grown under similar conditions. Data from these field trials are shown in table 2. The last item in each comparison is the traditional variety.

Production Increases

In 1968/69, the new rice varieties occupied about 7 percent (5.22 $\div$ 77.3) of the total rice area in South and Southeast Asia (table 1). As shown in table 2, the yield advantage of the new varieties seems to fall within the wide range of 30-100 percent. However, this advantage is in relation to rice raised under some of the best conditions--better irrigation and farmer skills--of the region. Average rice yields in the region before the introduction of the new varieties were about 1.6 metric tons per hectare. Table 2 suggests that under some of the better conditions yields may have been double the average. Thus, 7 percent of the area may have already been producing 14 percent of the rice. A doubling of yields (100-percent yield advantage) by use of new varieties would add another 14 percent to the output. On the other hand, a 30-percent yield advantage would add only 4.2 percent (30 percent times 14 percent). An average of this range would be about 9 percent, a rough estimate of the contribution of the new varieties to rice production, under normal weather conditions, on the area planted to these varieties in 1968/69. This is not an estimate of a growth rate.

Wheat is not nearly as important as rice in the less developed Asian countries, but as shown in table 1, about 16 percent (5.88 $\div$ 36.3) of the area was scheduled to be planted to new varieties of wheat in 1968/69. Wheat yields in the region averaged about 1 metric ton per hectare before the introduction of

Table 2.--Yield advantages of new varieties of rice and wheat, compared with traditional varieties, Asia, selected years

Variety	Yield per hectare	Remarks
Rice:	Metric tons	
Dwarf Indica	4.1	Both at 100 kilograms of nitrogen per hectare uniform variety trials, kharif, 1966, India (15, p. 8).
Local Indica	3.2	
IR-8	5.1	All at optimum marginal benefit-cost fertilizer application, experimental, wet season 1966 and 1967, IRRI, Los Banos, Philippines (5, p. 4).
IR-5	4.9	
Peta	2.7	
IR-8	6.2	Same conditions as above, Maligaya, Philippines (5, p. 4).
Peta	4.2	
IR-8	6.8	Same conditions as above except dry season, Maligaya, Philippines (5, p. 4).
IR-5	7.1	
Peta	4.0	
IR-8	n.a.	IR-8 showed 30 percent yield advantage over Peta (5, p. 11).
Peta	n.a.	
Wheat:		
Sonora 63	n.a.	Sonora 63 has a 30-35 percent yield advantage over local varieties, both at optimum levels of fertilization, based on experimental results in India (32, p. 695).
Local	n.a.	
Mexican	4.7	The 1966/67 crop in Ludhiana District, Punjab State, India. Mexican wheat was planted on only 11 percent of the wheat area in the district, probably by the best farmers on the best land (15, p. 9).
Indian	2.4	
Lerma Rojo		Tests and demonstrations, 1966/67-1967/68, India (3, p. 121).
Wheat 64 A	3.7-5.0	
Local	1.9-2.5	
Semi-dwarf	2.8	Both varieties grown on same farm, India (18).
Local	1.6	
"New"	1.8	Both irrigated, West Pakistan (18).
Local	1.0	

n.a. - not available

the new varieties. Again, table 2 suggests that yields under better conditions may have been twice that amount. Using the procedure followed above for rice, it can be roughly estimated that the new varieties would add from about 9.6 to 32 percent to normal wheat production in the region in 1968/69. The average of this range is about 20 percent.

This paper includes no forecast of future increases in production from the new varieties. Unless irrigation systems are extended, plantings will tend to be limited to the areas now adequately irrigated. Expansion on poorer land will tend to lower yields, but experience and adaptive research should help to overcome this problem. Farmer prices for grains, for competing products, and for inputs will affect both acreage and yields, but there is little information available on these factors for making estimates.

MULTIPLE CROPPING

The above estimates do not include the contribution from multiple cropping. An important characteristic of the new varieties is their shorter growing period and consequent potential for multiple cropping. However, even with good irrigation systems, multiple cropping requires a high level of managerial skill to coordinate a series of complex activities; hence, it is unlikely that it will spread quickly to areas where it is not already practiced. 4/ Multiple cropping may also create new problems. For example, in Thailand, in an area where year-round cropping has expanded, ''the presence of lush young plants throughout the year has not permitted the normal insect depletion common when the land was barren six months annually.'' This contributed to the spread of insect-borne disease (13). According to a recent survey, the potential land for double cropping of rice under existing irrigation is less than 10 percent of the total rice area in South and Southeast Asia, where the present double-cropped area amounts to 5 percent of the land in rice (3, p. 65). As irrigation systems are improved and extended and farmers gain experience, the area that is double cropped should be expanded.

In India, about 13 percent (18.6 million hectares) of the net area sown to all crops (138 million hectares) was double cropped in 1967/68. However, the bulk of the double-cropped area (12.5 million hectares) is unirrigated and thus unsuitable for high-yielding grain varieties. By 1969/70, it is expected that the irrigated double-cropped area will increase by only about 2 million hectares --less than 2 percent of the new sown area (33, p. D-5, D-23).

4/ Malaysia, a small producer of rice, seems to be an exception. Less than 5 years ago, an insignificant area was double cropped in Malaysia. In the 1968/69 season, 85,000 hectares were planted with a second crop of rice, compared with 40,000 hectares 2 years before. The area single cropped to rice amounts to about 325,000 hectares. Thus, more than 20 percent of the rice area is now double cropped (12).

LIMITATIONS TO SPREAD IN INDIVIDUAL COUNTRIES

Irrigation

A shortage of good irrigation systems appears to be the most important input limitation to the spread of the new varieties. Unless water can be carefully controlled, the advantage of the new varieties decreases rapidly. Many of the irrigation systems in South and Southeast Asia are not suitable for full realization of the potential of the new varieties. In many existing systems, the water flows by gravity from one field to the next, and fertilizer and plant protection chemicals are carried off in the water. Also, it is sometimes not possible to let the upper fields dry out in time for the harvest of the new varieties and thus the problem of wet grain at harvest is accentuated.

In much of Southeast Asia, the broad valleys will require large dams and long irrigation canals if additional irrigation systems are to be built (16, p. 339). Such systems cannot be built by local enterprise alone. Government action may be required to supply the initative, capital, and expertise, and new forms of cooperative organizations may be necessary to coordinate the use of the water.

In India and Pakistan, irrigation by pumps has grown rapidly. In West Pakistan, nearly 32,000 tube wells were installed by private enterprise in 5 years (3, p. 617). In many areas of Asia, there probably are large underground water resources which could be developed effectively. However, surveys and careful attention to water management will be necessary (3, p. 638).

As shown in table 3, only about 8 percent of India's grain area was planned to be under high-yielding varieties in 1968/69; therefore, it would seem that there is ample room for expansion. However, the 9.1 million hectares planned for high-yielding varieties represents 27 percent of the total irrigated grain area and a much higher, although undetermined, percentage of the land with reliable water control during the dry season. The data shown in table 3 support the view that inadequacy of irrigation is limiting the spread of the new grain varieties in India; the last column shows an almost equal increment in the planned acreage under high-yielding varieties and the planned increases in irrigated grain acreage. The irrigated area projected for traditional varieties remains very large and virtually stable. This suggests that water control on this area is not sufficiently reliable to risk the high costs of the fertilizer and insecticides required by the new varieties. No shortage of seed of the new varieties or of fertilizer is expected. In fact, some of the fertilizer available for 1969/70 will probably be used on traditional varieties.

West Pakistan has a good environment for the new wheat varieties--adequate irrigation, low rainfall, abundant sunshine, and few insects (5, p. 34). Long-standing problems of poor drainage and salinity are now being attacked. About 20 percent of West Pakistan's wheatland was planted to new varieties in the fall of 1968.

The potential for new varieties of rice in Pakistan is far more limited than for wheat. In East Pakistan, where 90 percent of Pakistan's rice is grown, the regular uncontrolled flooding of most of the producing areas lessens the value of the new short-stemmed varieties in the main spring and summer seasons

Table 3.--Grain production and inputs in India, 1967/68, and projections to 1968/69-1969/70

Item	Unit	1967/68	Projected 1968/69	Projected 1969/70	Change 1967/68-1969/70
Grain production	Million tons	1/ 100	98	103	2/ +10
Total grain area	Million ha.	1/ 121	119	119	2/ + 1
Irrigated grain area	do.	31.0	33.3	35.8	+ 4.8
High-yielding varieties of grain on irrigated land	do.	6.5	3/ 9.1	3/ 10.9	+ 4.4
Traditional varieties of grain on irrigated land	do.	24.5	24.2	24.9	+ 0.4
Fertilizer applied to grain 4/	Million nutrient tons	1.18	5/ 1.53	5/ 1.90	+ .72

1/ ''Normal weather'' estimate is 93.1 million tons and 118 million hectares.
2/ Relative to ''normal weather'' estimate.
3/ Plan of the Indian Government.
4/ Fertilizer applied to both high-yielding and traditional grain varieties.
5/ Requirements.

Source: (33, pp. D-4, D-16, D-17, D-20, D-22, D-24, D-26, D-55, E-1.)

(11, p. 19). In addition, insect and disease problems complicate the growing of new varieties in East Pakistan. West Pakistan, which has better growing conditions, produces basmati rice, an extra-long-grain variety. A significant share of this rice is exported at premium prices, and the government has increased the minimum purchase price to deter basmati producers from shifting to other varieties (25, p. 5). However, exports of basmati rice have declined rapidly in recent years.

The main obstacles to a rapid increase in the production of high-yielding rice in the Philippines appear to be the lack of good irrigation, a shortage of rice-drying facilities, and problems of consumer acceptability. The new varieties mature early during the latter part of the wet season. In 1967/68, because of a shortage of drying facilities, many farmers had to sell the new type rice wet in the fields at a 20-percent discount (4, p. 31).

Lack of water control and the inferior quality of the new rice varieties relative to export grades are deterrents to their spread in Thailand and Burma. A shortage of fertilizer at the farm level is a handicap in Burma and Indonesia (4, p. 31).

In Turkey, Mexican wheats seem to be adapted to the warmer coastal areas. The Turkish program for the expansion of acreage in these wheats developed very rapidly with little preparation (2, p. 19). In 1968/69, Mexican wheats were planted on about 7 percent of the total wheatland. It is likely that within a few years the Mexican seed will be grown on much of the southern and western coastal wheatlands, or on about 15 percent of Turkey's total wheat acreage.

New dryland wheat varieties may hold promise for increased yields in Turkey. Varieties are available which, under the proper conditions, greatly increase production without the necessity of irrigation. However, efficient dryland wheat farming is complicated and requires mechanization for proper tillage. The stubble-mulch system, which has been developed in the U.S. Great Plains, requires heavy equipment for subsurface plowing and deep planting. Introduction of such methods in areas of peasant farming will require the development of institutions to obtain and coordinate the use of heavy equipment.

Risks of Diseases and Pests

The new varieties of wheat and rice are exotic to most of the regions where they are being introduced and may become susceptible to local diseases and insect damage. In some countries, little adaptive research and testing was done before the new varieties were introduced. It is possible that microorganisms that were previously unimportant will become major causes of disease as field micro-climates are altered by heavy fertilization and the denser plant population of the new varieties. In the past, the use of locally produced seed of many strains provided some protection against the spread of diseases, since some of the various strains were resistant. The rapid introduction of a single variety on large contiguous areas increases the danger of epidemics (34, p. 468). Plant protection services, which require a high level of technical skill, are primitive in many of the less developed countries. Some problems with diseases and pests have already arisen. In India, the new rice varieties, which have spread much slower than the new wheats, have nonetheless been troubled more by insects and disease (1, rice, pp. 26-31; wheat, p. 26).

Prices and Incentives

The new varieties of rice sell at substantially lower prices than the traditional ones, since consumers generally do not rate them very highly (10, p. 43). The new rice generally is also considered inferior in milling qualities (9, pp. 7, 36). These characteristics, however, may be bred out in a few more years (4, p. 31). As indicated earlier, wet grain may often be a problem with the new rice varieties, but this disadvantage can presumably be overcome with investment in driers. Facilities for storage, processing, and transportation will be taxed as output increases. However, the private trade probably has considerable capacity for expansion to handle the increased production of grain (20, pp. vi, 22, 23).

The new varieties are exceptionally productive only when combined with fertilizer and pesticides, which the farmers must purchase. Thus, farmers using these inputs will of necessity be integrated into the markets; when prices for grain decline, they will tend to buy less fertilizer and pesticides. Of course, the prices of inputs are important, and recently the cost of fertilizer has declined. In the poorer countries, the demand for food grains generally is more elastic than it is in the richer countries, and thus substantial increases in production may be absorbed with relatively small price declines, if the problems of distribution are solved (20, p. 37). However, as the immediate food crisis abates, pressure for government investment in marketing facilities as well as in new irrigation may lessen.

In some cases, the margin of price over cost seems to be so large that considerable price declines could be absorbed without forcing farmers out of production and perhaps without even requiring much cutback in fertilization. Lower wheat prices in India do not seem to have slowed the spread of the new varieties (20, p. 54). However, these effects will depend on the alternatives available to farmers. Some will undoubtedly shift from grain to the production of other products if grain prices decline. In the Philippines, after the price of IR-8 fell, some farmers shifted to Malagkit, a high-priced rice used for cakes and pastries.

There is a rapidly growing demand for fruits, vegetables, and livestock products in the poorer countries. A recent study of the outlook for farm commodities in India projects that demand for milk and milk products will grow faster than supply during the next decade (21, tables 37-39). It is likely that some farmers located close to rapidly growing cities will shift to the production of fresh fruits and vegetables as grain prices decline. If grain prices decline sufficiently, more low quality rice and wheat probably will be fed to livestock. Such changes would be facilitated by research and investment to make inputs more productive in raising fruits and vegetables and in feeding livestock (20, p. 39).

THE BENEFICIARIES—THE PEOPLE LEFT BEHIND

Technological Innovations

The development of more productive varieties of grain is a technological change which lowers the unit costs of production and increases supplies available to consumers. In economic terminology, the new varieties cause a shift in the supply function; that is, at each price a greater amount of grain can be profitably produced and offered for sale than formerly. Technological innovations are among the main factors in economic development, and are important reasons why Malthus' predictions about population have not, and probably never will, come true. On the other hand, technological developments are not magic, and while solving some problems or providing opportunities to solve them, they often create other major economic, social, and political problems of adjustment (31, pp. 77-87; 34, p. 475).

Technological developments affect both outputs and inputs. The effects on outputs are usually desirable; the production of increased amounts of socially desirable goods and the reduction in costs give rise to the possibilities of increased welfare and a lower cost of living. To a large extent the benefits of technological development in agriculture are soon widely distributed among consumers, at least among those consumers who have the purchasing power and the opportunity to take advantage of lower market prices. Lower costs of living may be reflected in lower labor costs and thus have pervasive effects on overall economic development (19, pp. 95, 96). The increased production also may substitute for costly imports. However, in evaluating the effects of the new grain varieties in the poorer countries, even these conclusions must be qualified. A judgment must be made whether the increased output can be expected to result in increased total supplies to consumers or whether it may to a considerable extent displace grain obtained cheaply under concessional arrangements. In some countries grain prices have been high and subsidies substantial, a condition which may have stimulated some high-cost, uneconomic production (20, pp. 55, 56).

The People Left Behind

Not all the effects of technological changes on inputs may be desirable. The owners of some inputs will find that the demand for their services is lessened, at least in their present uses.

The development of the new grain varieties implies a shift in the comparative advantage among areas producing grain. The irrigated areas most suitable for these new varieties gain an advantage relative to other grain-producing areas, and the owners of suitable resources will be benefited. To a considerable extent, unless taxed away, the benefits of these developments will go to owners of irrigated land. Those farmers who are early users of the new process will tend to benefit from increased production and extra income until competition lowers profits. Thus, those farmers who are in a position (because of irrigation, location, credit availabilities, knowledge, etc.) to take advantage of the new opportunities may do very well.

Those farmers who are not in such an advantageous position will find that the increased competition reduces their market opportunities and thus causes them additional difficulties. Opportunities for some of these farmers might be found by research directed at increasing the productivity of labor in dairying and in growing fruits and vegetables (20, p. vi). Policies to improve the operation of the labor market and increase the demand for agricultural labor should have high priority. It is almost universally true in the less developed countries that the nonfarm sector is not growing rapidly enough to absorb a large influx of displaced farm labor. In most of these countries, the labor force in agriculture will continue to increase for some decades. Thus, a strategy for agricultural development which will promote labor-absorbing activities will be important. It has been suggested that the labor-using, capital-saving approach to agricultural development followed in Japan and Taiwan provides a model which should be studied in relation to the ''seed-fertilizer revolution'' (17, p. 2).

The tendency for certain groups to be especially benefited may be accentuated in the adoption of new varieties in some countries because resources are concentrated in subsidized ''packages'' under government programs. There has been a tendency to make the packages available to those regions or farmers where the production response is likely to be greatest. Because of the complementarity between the various inputs required (seeds, chemicals, equipment, credit, information, storage, and processing), the ''package'' approach has been helpful in the rapid spread of the new varieties. However, these programs have been so profitable for the adopting farmers that some of the subsidies used may have been unnecessary. Even in cases where subsidies helped to obtain rapid initial adoption, they will be less necessary as farmers become aware of the potential of the new seed combined with fertilizer. Pakistan, for example, has reduced its fertilizer subsidy (20, p. 56).

In some areas, the success with the new grain varieties has already stimulated considerable investment in machinery, but a lack of machinery, especially drills, threshers, and land leveling equipment has held down yields (18, pp. 13, 14).

Many of the developed countries now have serious social and economic problems arising from the displacement of farm people by technological developments. However, the developed countries have had rapid growth in their nonfarm sectors to absorb most of such people and have been able to afford high welfare costs for some of the others. The developed countries have also had relatively good labor markets and high levels of education which tend to make labor transferable.

The implications of the above arguments are not that technological development is undesirable. Technological changes are inevitable and essential for economic development. However, a few, very limited technological changes cannot be expected by themselves to give a great impetus to economic development. On the other hand, a particular technological change may generate considerable social dislocation and even discord (22, p. 12). The poorer countries do not have all the options open to the developed countries in solving the problems caused by the uneven impact of technological changes. To ignore the existence of these problems might be disastrous. Therefore, it is imperative that close study be given to all alternatives. Research to evaluate these problems and develop policies to solve them should have a high priority.

THE OUTLOOK FOR WORLD SUPPLY AND DEMAND

Demand—Self-Sufficiency

Several recent studies suggest that production of food grains in some countries considered in this report may increase at a rate of 4 to 6 percent a year (11, pp. 19, 27-28; 15, p. 13). On the other hand, it is unlikely that effective economic demand for grain in any of these countries will increase faster than 4 percent a year (less than 3 percent for population growth and perhaps 1 percent from rising per capita income), unless the livestock industry can be developed fast enough to use substantial amounts for feed. Countries now importing grain may use increased domestic production to replace imports. As a precentage of total consumption, imports of grain in most less developed countries are relatively small. Thus, the growth of production at a faster rate than demand could soon eliminate the need for imports. It seems possible that the less developed countries of Asia can generally become self-sufficient in grain before many years, although imports may continue to supply some large coastal cities. Turkey and the Philippines are nearly self-sufficient in food grains. Both Pakistan and India plan to be self-sufficient within a few years.

Self-sufficiency is not necessarily a wise economic goal, however. To the extent that imports of grain are paid for with scarce foreign exchange, the concern for self-sufficiency is understandable. Of course, most of the imports of grain by these countries have been made as food aid with large price discounts. Still, there may be a desire to be free from the uncertainty and restrictions of large imports of food aid. The tightening of the conditions under which U.S. food aid is given undoubtedly reinforces this desire.

As indicated earlier, the demand for grains in the poorer countries probably is elastic enough so that prices need not decline greatly to bring substantial increases in consumption. It has been suggested that, because of the interrelations between agricultural output, income, and demand, ''. . . in early stages of development, increasing agricultural production will not have a strong effect upon prices'' (19, p. 76). However, this conclusion may not be correct in the case of large increases of production in concentrated areas and in poorly organized markets; a description which fits many of the situations wherein the increases of the new varieties have occurred. Despite surpluses of grain in West Pakistan, limited transportation and handling facilities restricted that country's ability to supply East Pakistan in the spring of 1969. In any case, after reaching self-sufficiency, if supplies of grain grow faster than domestic demand, prices will be depressed to some degree, unless the international market can absorb additional increases.

Shortrun Outlook for Grain Exports of the Less Developed Countries

Current world supplies of grain are large relative to demand. In the shortrun, this relationship probably will continue, barring widespread drought or radical changes in policy. In many countries, producer prices are well above export prices; the developed countries in this category are moving grain in world trade with the aid of subsidies. It is unlikely that the less developed countries of Asia will be able to afford to enter into large-scale competition in this market.

At present, Thailand stands almost alone among less developed Asian countries as a major world exporter of grain. The bulk of Thailand's grain is produced at relatively low cost with modest cash inputs and, consequently, is exported without subsidy. On the other hand, several Asian countries producing the new grain varieties have already encountered problems disposing of surpluses. High producer prices have been incentives to the adoption of the new grain technology. An alternative to disincentive prices would be to keep producer prices relatively high and subsidize exports. But the gap between producer and export prices in many Asian countries is wide, and large outlays might be required.

Here, the experience in Mexico may be relevant. Mexico pioneered in adoption of the wheat varieties now being introduced in Asia. In 1963/64, when Mexico became a substantial net exporter of wheat, world wheat prices were at a peak because of the short crop in the USSR. The f.o.b. export price of Mexican wheat was only about 5 percent below the Mexican producer price. In subsequent years, however, the gap widened substantially as export prices fell and Mexico was burdened with a large export subsidy. Since 1966, the Mexican Government has altered its price policy to favor a shift in acreage from surplus wheat to those crops in short supply (such as sorghum and oilseeds). However, the 1966 reduction in wheat support was not sufficient to bring domestic prices down to export price levels.

The Longer Run Outlook for International Grain Markets

The high prices which held in the international rice market in 1967-68 may well be things of the past, and it seems likely that actual surpluses of wheat, or the capacity to produce surplus wheat, will be bearing down on prices for some time to come. In world markets, prices of food grains have tended to be higher than those for feed grains, partly because of government programs specifically designed to benefit food grains. Increased supplies of food grains coming from the new developments will increase the pressure on prices of food grains relative to feed grains.

Even for rice and feed grains, the surplus problem, although less acute than for wheat, may also grow. Thus, market pressures will tend to depress the prices of grains (especially wheat) from recent levels unless offset by government actions. The actual balancing of supply and demand may be made by various combinations of government programs and market forces.

A major difficulty in forecasting future world grain markets is that the policies of governments affect the world grain economy in many and complex ways through price support programs, subsidies, taxes, trade agreements, and food aid. Such government policies have a profound effect on production, utilization, and trade in grains. Although theoretically the effects of changing policies may be analyzed by economic models, the changes in policies themselves cannot be predicted by economic analysis. The economist must rely on ''assumptions,'' which often are of doubtful validity. Since the United States is the world's leading producer and exporter of grain and the largest donor of food aid and other assistance, its policy decisions are of primary importance in determining how world grain trade will actually develop.

During the next decade, assuming general political stability, standards of living and consumption of grain probably will rise throughout the world. In the less developed countries, the per capita growth of consumption in general, and of grains in particular, will be slow. But there is little evidence to support the view that in this period, discounting possible natural disasters, the world food situation will tend to worsen. On the other hand, there will still be many undernourished people in the world, people who are too poor to pay for adequate nutrition.

BIBLIOGRAPHY

(1) Agency for International Development
1969. New Cereal Varieties, Rice and Wheat in India. Spring Review of the New Cereal Varieties, March.

(2) ──────────
1969. The Role of Research. Spring Review of the New Cereal Varieties, May.

(3) Asian Development Bank
1968. Asian Agricultural Survey. Vol. II, Sectional Reports, Manila, March.

(4) Barker, Randolph
1968. The Role of the International Rice Research Institute in the Development and Dissemination of New Rice Varieties. International Rice Research Institute, Los Banos, Philippines, 47 pp.

(5) ──────────, Liao, S. H., and Datta, S. K.
1968. Economic Analysis of Rice Production from Experimental Results to Farmer Fields. Paper presented at Agronomy Department Seminar, University of the Philippines, Manila, Aug. 9.

(6) Brown, Lester R.
1968. The Agricultural Revolution in Asia. Foreign Affairs, Council on Foreign Relations, Inc., New York, July: 688-698.

(7) Cannon, Grant
1967. On the Eve of Abundance. The Farm Quarterly, Fall Forecast: 65.

(8) Chandler, Robert F.
1968. Dwarf Rice--A Giant in Tropical Asia. U.S. Dept. Agr. Yearbook (Science for Better Living), 1968.

(9) Corty, Floyd L.
1969. Global Crop Paper, Rice. Spring Review of the New Cereal Varieties, Agency for International Development, May.

(10) Dalrymple, Dana G.
1969. Technological Change in Agriculture, Effects and Implications for the Developing Nations. International Development, Foreign Agricultural Service, U.S. Dept. Agr., in cooperation with the Agency for International Development, Washington, D.C.

(11) Food and Agriculture Organization of the United Nations
1968. Reviews of Medium-Term Food Outlook. Committee on Commodity Problems, Forty-Third Session, CCP 68/16, Aug. 14.

(12) Foreign Agricultural Service
1968. Dispatch AGR-73, from the American Embassy, Kuala Lumpur, Oct.

(13) ──────────
1969. Dispatch AGR-9041, from the American Embassy, Bangkok, Thailand, March 24.

(14) ──────────
1968. Dispatch AGR-158, from the American Embassy, Rawalpindi, Pakistan, Aug. 7.

(15) Hendrix, W. E., Naive, J. J., Adams, W. E.
1968. Accelerating India's Food Grain Production 1967-68 to 1970-71. U.S. Dept. Agr. Foreign Agr. Econ. Rpt. 40.

(16) Hsieh, S. C., and Ruttan, V. W.
1968. Environmental, Technological, and Institutional Factors in the Growth of Rice Production: Philippines, Thailand, and Taiwan. Reprinted from Food Research Institute Studies, Vol. VII, No. 3, 1967, Stanford University, Stanford, Calif.

(17) Johnston, Bruce F. and Cownie, John
1969. The Seed-Fertilizer Revolution and the Labor Force Absorption Problem. Food Research Institute, Stanford University, Jan. 20, (unpublished).

(18) Kronstad, Warren E.
1969. Global Crop Paper, Wheat. Spring Review of the New Cereal Varieties, Agency for International Development, May.

(19) Mellor, John W.
1966. The Economics of Agricultural Development. Cornell University Press, Ithaca, New York.

(20) ──────────
1969. The Role of Government and the New Agricultural Technologies. Spring Review of the New Cereal Varieties, Agency for International Development, May.

(21) National Council of Applied Economic Research
1969. Supply of and Demand for Selected Agricultural Products in India, Revised Projections to 1980-81. New Delhi, April.

(22) New York Times
1969. Madras is Reaping a Bitter Harvest of Rural Terrorism, Jan. 15: 12.

(23) Organisation of Economic Co-operation and Development
1967. The Food Problem of Developing Countries. Paris, Dec.

(24) Paddock, William and Paul
1967. Famine 1975! America's Decision: Who Will Survive? Little, Brown and Company, Boston.

(25) Parker, John, Jr.
1968. Preview: Foodgrain Needs in Pakistan. Foreign Agriculture (weekly), Foreign Agricultural Service, U.S. Dept. Agr., Sept. 16.

(26) Reitz, Louis P.
1968. Short Wheats Stand Tall. U.S. Dept. Agr. Yearbook (Science for Better Living), 1968.

(27) Rice, E. B.
1969. The Role of Institutions. Spring Review of the New Cereal Varieties, Agency for International Development, May.

(28) Rosenthal, Jerry E.
1968. The Philippine Rice Story. War on Hunger, A Report from the Agency for International Development, Jan.: 4.

(29) Ruttan, Vernon W.
1968. Growth Stage Theories, Dual Economy Models and Agricultural Development Policy. Department of Agricultural Economics, University of Minnesota, AE 1968/2.

(30) Schertz, Lyle P.
1968. The Role of Farm Mechanization in the Developing Countries. Foreign Agriculture (weekly), Foreign Agricultural Service, U.S. Dept. Agr., Nov. 25.

(31) Schultz, Theodore, W.
1965. Economic Crisis in World Agriculture. The University of Michigan Press, Ann Arbor.

(32) The President's Science Advisory Committee
1967. The World Food Problem. Report of the Panel on the World Food Supply, May.

(33) US/AID Mission.
1968. India Program Memorandum, FY 1970. New Delhi, Sept.

(34) Wharton, Clifton R., Jr.
1969. The Green Revolution: Cornucopia or Pandora's Box? Foreign Affairs, Council on Foreign Relations, Inc., New York, April: 464-476.

(35) Williams, Joseph R.
1967. Turkey's Crash Wheat Program. Foreign Agriculture (weekly), Foreign Agricultural Service, U.S. Dept. Agr., Nov. 20.

HIGH-YIELDING
VARIETIES OF WHEAT
IN DEVELOPING COUNTRIES

U.S. DEPARTMENT OF AGRICULTURE • Economic Research Service

ABSTRACT

The production of high-yielding varieties of wheats in developing countries in recent years has been so successful as to be termed a Green Revolution. These wheats (semidwarfs), originating in Mexico, are short stemmed, photo insensitive, and highly responsive to inputs. Total semidwarf wheat area in India, Pakistan, Mexico, Turkey, Afghanistan, Tunisia, Iran and Morocco, expanded rapidly from 0.6 million hectares in 1966 to 10.6 million hectares in 1970. Production during the same period increased from 1.6 million tons to 22.7 million tons. In 1970, semidwarf wheat in these countries accounted for 25 percent of the total wheat area and 49 percent of the total wheat output. This agricultural advance is attributed to programs developed by individual governments with the assistance of international institutions and private agencies.

Key words: High-yielding wheat varieties, semidwarf wheat, wheat production, Green Revolution, developing countries.

FOREWORD

The worldwide importance of the Green Revolution--the introduction and use of high-yielding varieties (HYV) of semidwarf wheat--was recently attested by the awarding of the Nobel Prize to Dr. Norman E. Borlaug, the developer of the new wheat varieties. Because this revolution in agriculture affects the economies of both the countries growing high-yielding varieties and their trading partners, it is important that all aspects of current semidwarf wheat cultivation be known and its potential be evaluated.

This report assesses the current situation with regard to the semidwarf wheat varieties developed in Mexico and the success of programs associated with their introduction and use outside of Mexico--in India, Pakistan, Turkey, and elsewhere. It also examines the relationships of wheat area, production, and yield.

In the compilation of this report, the author is indebted to numerous individuals in U.S. Agency for International Development (USAID) missions, U.S. Embassies, the International Maize and Wheat Improvement Center (CIMMYT), foreign Governments, private foundations, international agencies, and in Wash., D.C., who have volunteered their expertise in agronomy and the agricultural economics of the countries under study. The particular encouragement and assistance given the author of this work by James J. Naive deserves special note.

The choice of data and information for this report as well as the conclusions drawn are, of course, the sole responsibility of the author.

G. Stanley Brown, Chief
Europe and Soviet Union Branch
Foreign Regional Analysis Division

CONTENTS

	Page
SUMMARY	vi
INTRODUCTION	1
TRENDS AND DEVELOPMENTS	6
TECHNOLOGICAL EFFECTS OF HIGH-YIELDING VARIETIES	9
COUNTRY SITUATIONS	12
India	12
Pakistan	16
Turkey	20
Afghanistan	24
Tunisia	24
Iran	26
Morocco	26
Other countries	27
CIMMYT	28
CONCLUSIONS	28
LITERATURE CITED	30
APPENDIX TABLES	37

Washington, D.C. 20250   September 1971

TEXT TABLES

Table		Page
1.	Shipments of research germ plasm by CIMMYT to six world regions, 1968-69..	2
2.	Wheat: Area and production of high-yielding and local varieties, selected countries, annual, 1966-70............................	7
3.	India: Area, production, and yield of wheat; local and high-yielding varieties, annual, 1966-70............................	13
4.	Pakistan: Area, production, and yield of wheat; local and high-yielding varieties, annual, 1966-70.......................	17
5.	Turkey: Area, production, and yield of wheat; local and high-yielding varieties, annual, 1966-70.......................	21

FIGURES

Figure		Page
1.	LDC's growing HYV wheat, 1966-70 and world wheat production......	4
2.	Wheat sowing and harvesting seasons............................	5
3.	Wheat production..	10
4.	Hypothetical production functions for local and high-yielding varieties of wheat.....................................	11
5.	India: Wheat area, production, and yield, semidwarf and total...	14
6.	Pakistan: Wheat area, production, and yield, semidwarf and total...	18
7.	Turkey: Wheat area, production, and yield, semidwarf and total...	22

SUMMARY

The introduction and cultivation of newly developed high-yielding varieties of wheat (semidwarfs) in developing countries have been so successful in boosting wheat production as to be termed a Green Revolution.

In the period 1966-70, area given to the new varieties--chiefly by displacing local wheats--in India, Pakistan, Mexico, Turkey, Afghanistan, Tunisia, Iran, and Morocco expanded rapidly from 0.6 to 10.6 million hectares, while production rose spectacularly from 1.6 to 22.7 million tons. In 1970, semidwarf wheat in these countries accounted for 25 percent of the total wheat area and 49 percent of the total wheat output. In all countries, semidwarf yield per hectare greatly exceeded that of local varieties--often by two and three times. In 1970, high-yielding varieties averaged 2,140 kilos per hectare, compared with 770 kilos for local varieties. Average yields of all wheat have increased from 888 kilos per hectare in 1966 to 1,120 kilos in 1970.

Among the countries now using semidwarf wheat, results have been most dramatic in India where semidwarf production rose from 8,000 tons in 1966 to 13,606,000 tons in 1970, and in Pakistan where it increased from 14,000 to 4,350,000 tons in the same period. Because of adverse weather conditions, yields of these varieties in Turkey were less spectacular. But even here, semidwarf production rose from negligible amounts in 1966 to 1,797,000 tons in 1970.

On a worldwide basis, the 10.6 million hectares and 22.7 million tons of semidwarf production represent 5.2 and 7.9 percent of total wheat acreage and production, respectively.

Semidwarf wheats, developed in the early 1960's in Mexico by Nobel Peace Prize winner Dr. Norman E. Borlaug under the sponsorship of the Mexican Government and the Rockefeller Foundation, have the following characteristics: (1) Short and strong stems which reduce lodging and fallover; (2) greater response to fertilizer and water inputs; (3) photo-insensitivity which permits their cultivation in latitudes outside the normal growing belt; (4) earlier maturity which facilitates multiple cropping; and (5) good resistance to rusts and other diseases. A "package program" involving specific cultural practices has been developed to ensure successful cultivation of the high-yielding varieties.

HIGH-YIELDING VARIETIES OF WHEAT IN DEVELOPING COUNTRIES

by

Sheldon K. Tsu, Agricultural Economist
Foreign Regional Analysis Division, Economic Research Service

INTRODUCTION

The excellent performance of high-yielding semidwarf wheat in developing countries has been widely heralded as a major agrotechnological breakthrough. 1/ Because of the vigorous promotion of the wheat program and spontaneous response of the farmers, the high-yielding varieties (HYV) program has been termed a Green Revolution. Though the Green Revolution commonly refers to the introduction and successful use of high-yielding varieties of all grains, most of the acclaim has been given to the results of the semidwarf wheat varieties. 2/ This revolution in agriculture has contributed to a solution of the world food problem.

Although development of these varieties can be traced to the latter part of the 19th century, intensive research began only about 20 years ago. During this period many institutions, public and private, sponsored research on the wheat varieties and promoted their distribution to and use by farmers. These included the Rockefeller and Ford Foundations, the International Maize and Wheat Improvement Center (CIMMYT), the U.S. Department of Agriculture (USDA), the U.S. Agency for International Development (USAID), the Food and Agriculture Organization of the United Nations (FAO), and many universities and research institutions 3/ in both developing and developed countries. From the standpoint of international coordination of varietal improvement programs and the collection and distribution of seed stock (germ plasm), during 1968-69, CIMMYT distributed 146 shipments of germ plasm to 61 countries throughout the world (table 1).

The semidwarf wheat varieties that have been particularly successful in Mexico, India, and Pakistan were developed by Dr. Norman E. Borlaug and colleagues at the National College of Agriculture in Chapingo, Mexico, under

---

1/ For further discussion of this development, see (4, 6, 7, 8, 13, 24). Underscored numbers in parenthesis refer to Literature Cited, p. 30.

2/ HYV rice varieties developed at the International Rice Research Institute at Los Banos, Philippines, have only had moderate success thus far (46), (95).

3/ See (24) for a fuller discussion of these institutions.

Table 1.--Shipments of research germ plasm by CIMMYT to six world regions, 1968-69

Germ plasm type	North America	South America	Near and Mid East	Europe	Africa	Others 1/	Countries receiving shipments
International spring wheat yield nursery	2	6	11	6	10	4	39
Commercial varieties	1	4	--	9	6	1	21
Advanced lines	2	3	3	2	--	1	11
F2	1	3	3	1	5	--	13
F3	1	3	1	3	2	3	13
International nursery for durum	1	2	6	3	3	2	17
International nursery for triticale	2	3	3	5	3	1	17
International screening nursery	1	4	3	1	4	2	15
Total shipments	11	28	30	30	33	14	146

1/ Includes USSR, Oceania, and the Far East

Source: (8), p. 81.

the joint sponsorship of the Mexican Government and the Rockefeller Foundation. 4/ These semidwarf wheats, which are generally less than 80 centimeters (32 inches) in height, are highly responsive to fertilizers when water supplies are adequate. They are generally classed as soft spring wheats. Because of their spring habit, the Mexican varieties cannot be raised in colder areas where conditions would require a period of dormancy. Having a short stiff straw, the semidwarf varieties 5/ show virtually no tendency to lodge or fall over. Photo-insensitivity is the characteristic which

---

4/ See (24, 40, 41) for a full discussion.
5/ Hereafter, the terms "semidwarf wheat", "HYV", and "Mexican wheat" will be used synonymously.

enabled Mexican wheat to be grown successfully outside of Mexico. Because of the insensitivity to length of daylight, they can be grown successfully within a wider latitudinal range throughout the world. Other high-yielding semidwarf varieties have been developed over the years in various countries, including the United States, but they have been unsuccessful when taken out of their natural habitat.

Experiments as well as experience show that for farmers to realize the potential of the HYV, the seeds must be used in conjunction with certain cultural practices. This recommended combination of seed use and cultural practices is called a "package program" 6/ The package program includes such cultural practices as adequate seedbed preparation; sufficient inputs of fertilizer, water (irrigation), and pesticides; and use of right type of implements. Without adequate application of inputs, for instance, HYV performance would not differ significantly from that of local varieties.

At the same time, governments have strengthened the agricultural infrastructure by improving farm extension service as well as marketing and distribution systems, and by implementing favorable farm price and subsidy policies.

This report focuses on the extent to which the Mexican varieties and their derivatives have contributed to total wheat production in developing countries. Particular attention is given to the 1970 situation and--where information was available--the outlook for 1971. In addition, activities on the current status of semidwarf improvements or breeding projects are included. The intent of the report is to provide a better basis for evaluating the results of HYV programs.

Area coverage (see fig. 1) here includes all developing nations that have taken part in a semidwarf program, although emphasis is given to those that have been most active in sowing semidwarf varieties. 7/ Data on area and production of 1969 and 1970 are preliminary and subject to revision.

Sowing and harvesting seasons are shown in figure 2. For the most part, the wheat crop in the countries under review is sown in the fall and harvested in the spring of the following calendar year. Hereafter in this report, data on area, production, and yield will refer to the year in which the crop was harvested. Metric units are used unless noted otherwise.

6/ In India, a similar program for general agriculture was initiated in 1961 as the "Intensive Agricultural District Program (IADP)" by the Ford Foundation on the recommendation of a study group headed by Dr. Sherman Johnson.

7/ The discussions will center on eight countries--India, Pakistan, Mexico, Turkey, Afghanistan, Tunisia, Iran, and Morocco. In addition, other countries now in the testing and trial stage include Nepal, Burma, Syria, Jordan, Lebanon, Iraq, Israel, Algeria, Egypt, South Africa, Rhodesia, Kenya, Sudan, Saudi Arabia, Tanzania, Zambia, Argentina, Brazil, Bolivia, Colombia, Guatemala, and Paraguay.

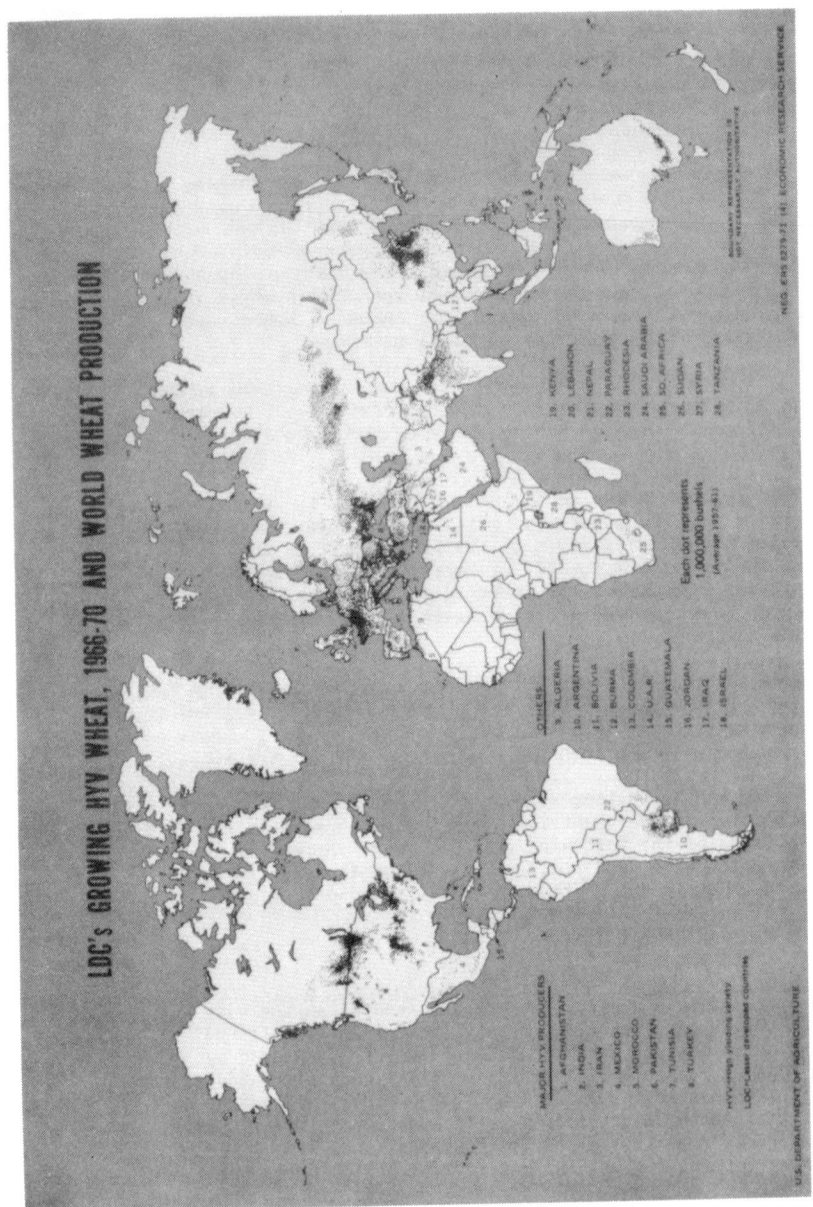

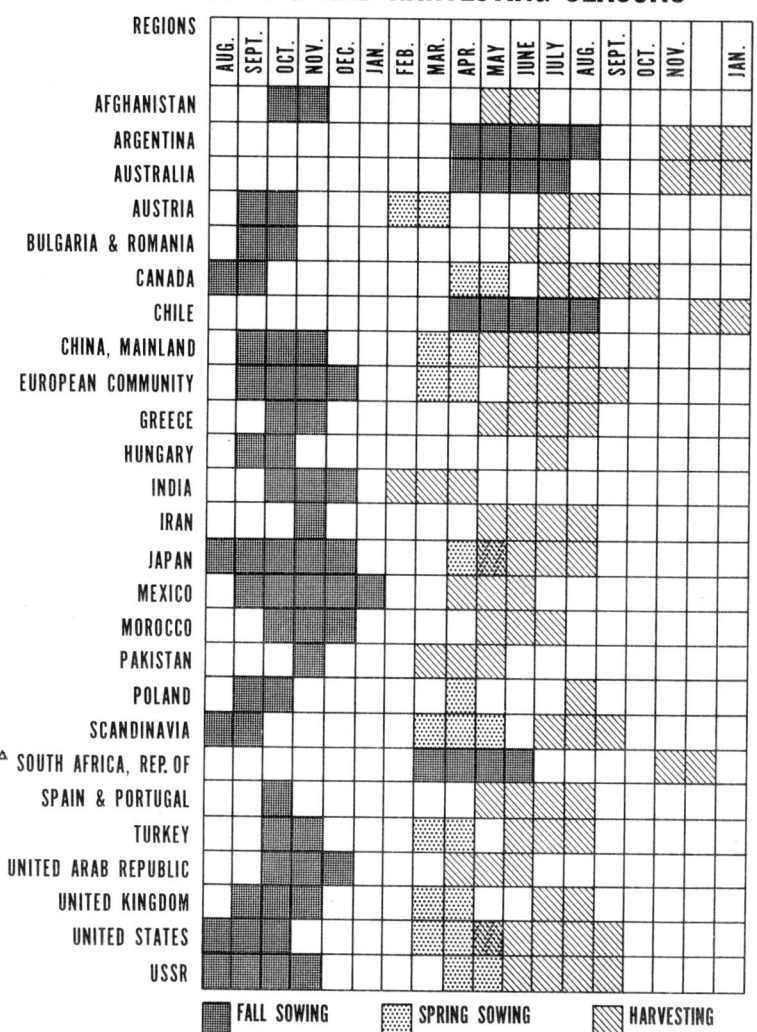

TRENDS AND DEVELOPMENTS

The expansion of semidwarf wheat area in the countries under study since 1966, when they were first grown outside of Mexico on a significant scale, has been spectacular. Area under cultivation increased from 0.6 million hectares in 1966 to 10.6 million hectares in 1970 (table 2). Most of the expansion occurred on irrigated land. The successful performance of these high-yielding varieties in 1966 and 1967 coupled with the dire need for accelerating food grain output in India and Pakistan spurred a tremendous area expansion in 1968 and 1969. The rate of expansion slowed somewhat in 1970 as available irrigated land decreased.

The area devoted to local varieties has declined over 8 percent since 1966, but the total wheat area has increased 21 percent. This indicates that the semidwarfs have not only replaced local wheats but also expanded into new areas. There is evidence that new land has been brought into cultivation with the expansion of irrigation and that semidwarfs have been substituted for other crops. Photo-insensitivity has permitted the HYV's to be grown in areas outside of the normal wheat belt. In addition, part of the gross area increase is due to more multicropping because the high-yielding varieties mature earlier than the indigenous varieties.

On a worldwide basis in 1970, the 10.6 million hectares and 22.7 million tons of semidwarf wheat in the countries under study were 5.2 and 7.9 percent of the total wheat area and production, respectively (32). These relationships are heavily influenced by the performance of Mexico, India, and Pakistan, which accounted for 80 percent of the 1970 total area and output of the eight countries. Turkey also has relatively large semidwarf areas (see appendix table 4).

All reports indicate that yields of these varieties markedly exceed those of the indigenous varieties. But yields fluctuate year after year primarily because of weather and input variations. The replacement of local varieties by semidwarf wheats appears to be the prevailing pattern. During the test plot and experimental stage, HYV yields are extremely high because sufficient inputs, proper cultivation, and better management are exercised in a limited area under appropriate control for growing. Initially, the semidwarf are sown on the most fertile lands, which are kept well irrigated and heavily fertilized.

However, yields tend to decline during the following stage. The main cause of this paradoxical situation is the rapid expansion of the HYV growing areas, particularly in countries where such sudden area increases take place under "crash programs." It is obvious that during this transitional period, insufficient supply and application of inputs plus the inexperience of farmers may adversely affect yields. Lands of lower fertility are gradually brought into the semidwarf wheat cultivation which would also limit the yield capacity. However, after this interim slippage period, the programmed activities are strengthened and HYV yields increase.

Table 2.--Wheat: Area and production of high-yielding and local varieties, selected countries, annual, 1966-70

Country and varieties		1966 Area 1,000 ha.	1966 Production 1,000 tons	1967 Area 1,000 ha.	1967 Production 1,000 tons	1968 Area 1,000 ha.	1968 Production 1,000 tons	1969 Area 1,000 ha.	1969 Production 1,000 tons	1970 Area 1,000 ha.	1970 Production 1,000 tons
India	Local	12,653	10,416	12,324	10,115	12,056	8,697	11,165	7,460	10,515	6,487
	HYV	3	1/ 8	514	2/1,278	2,942	2/7,843	4,793	3/11,191	6,111	3/13,606
	Total	12,656	10,424	12,838	11,393	14,998	16,540	15,958	18,651	16,626	4/20,093
Pakistan	Local	5,205	3,938	5,316	4,170	5,104	4,240	3,657	2,762	3,516	3,049
	HYV	5	14	101	224	957	2,237	2,388	3,949	2,833	5/4,350
	Total	5,210	3,952	5,417	4,394	6,061	6,477	6,045	6,711	6/ 6,349	5/7,399
Mexico	Local	---	---	---	---	---	---	---	---	---	---
	HYV	635	1,609	762	2,057	717	1,793	715	2,000	715	2,250
	Total	635	1,609	762	2,057	717	1,793	715	2,000	7/715	7/2,250
Turkey	Local	7,163	8,200	7,203.4	8,998	7,134	7,805	7,071	6,672	6,937	6,203
	HYV	n.a.	n.a.	0.6	2	170	8/595	579	8/1,628	623	8/1,797
	Total	7,163	8,200	7,204	9,000	7,304	8,400	9/7,650	9/8,300	9/7,560	9/8,000
Afghanistan	Local	2,280	2,033	2,364.2	2,547	2,350	2,594.2	2,745	2,490	2,819	2,291
	HYV	n.a.	n.a.	1.8	3	22	65.8	10/122	10/289	10/147	10/371
	Total	2,280	2,033	2,366	2,550	2,372	2,660	10/2,867	10/2,779	10/2,966	10/2,662
Tunisia	Local	845	349	815	282	649.2	381.5	627	326	697	330
	HYV	n.a.	n.a.	n.a.	n.a.	0.8	1.5	13	11/24	53	12/120
	Total	845	349	815	282	650	383	640	350	12/750	12/450
Iran	Local	4,000	3,190	4,400	4,000	4,800	4,400	4,590	3,884	4,610	3,680
	HYV	n.a.	n.a.	n.a.	n.a.	n.a.	n.a.	10	13/16	90	13/120
	Total	4,000	3,190	4,400	4,000	4,800	4,400	14/4,600	14/3,900	14/4,700	14/3,800
Morocco	Local	1,637	814	1,777	1,090	1,977	2,410	1,759	1,606	1,869	1,850
	HYV	n.a.	n.a.	n.a.	n.a.	0.2	15/0.3	5	15/7	15/10	16/20
	Total	1,637	814	1,777	1,090	1,977.2	2,410.3	17/1,764	17/1,613	17/1,879	17/1,870
Total	Local	33,783	28,940	34,199.6	31,202	34,070.2	30,527.7	31,614	25,200	30,963	23,890
	HYV	643	1,631	1,379.4	3,564	4,809	12,535.6	8,625	19,104	10,582	22,634
	Total	34,426	30,571	35,579	34,766	38,879.2	43,063.3	40,239	44,304	41,545	46,524

Note: n.a. indicates not applicable or not available.

Sources: (over).

Sources: Data on total area and production 1966-70 (footnote 18, below); HYV area data (12) except where footnoted; high-yielding variety production data (6, 7, 8) except where footnoted:

1/ (6); Letter from James H. Boulware, Agricultural Attache, American Embassy, New Delhi, India, Jan. 1970.
2/ (35)
3/ USDA estimates (m.a.)*
4/ Agricultural Attache Office
    1971. Agricultural Situation, American Embassy, New Delhi, India, Jan.
5/ _____
    1970. Carl O. Winberg, Agricultural Attache, American Embassy, Rawalpindi, Pakistan, June.
6/ _____
    1971. Grain and Feed Report, PK 1011, American Embassy, Islamabad, Pakistan, Feb.
7/ _____
    1970. Annual Grain and Feed Report, MX0071, American Embassy, Mexico, Nov.
8/ _____
    1970. Harvey R. Varney, Agricultural Attache, Embassy, Ankara, Turkey, June.
9/ _____
    1970. Grain and Feed, TRO-076 American Embassy, Ankara, Turkey, Nov.
10/ _____
    1970. RGA Request for PL 480 Wheat to AID, A-375, Joint American Embassy/USAID, Kabul, Afghanistan, Sept.
11/ (8) with USDA estimates (m.a.)*
12/ Agricultural Attache Office
    1970. Agricultural Report, IN-0002, American Embassy, Rabat, Morocco, Dec.
13/ (8) with USDA estimates (m.a.)*
14/ Agricultural Attache Office
    1971. Agricultural Situation, IR-1001, American Embassy, Tehran, Iran, Jan.
15/ (70, 71, 72)
16/ Letter from Ralph J. Edwards, AID Cereal Project Manager, Rabat, Morocco, July, 1970.
17/ Agricultural Attache Office
    1971. Grain and Feed, MO 1006, American Embassy, Rabat, Morocco, Feb.
18/ Foreign Regional Analysis Division
    1971. ERS Crop Data Base, U.S. Dept. Agr., Apr.

* methodology available.

The widespread acceptance of semidwarf varieties by growers during the period under review has been encouraging. Traditional cultural practices were apparently no obstacle to farmers' acceptance once the profitability of semidwarf production became evident. Government policies for both inputs and outputs have had an important role in providing economic incentives for cultivators. Information on these varieties was disseminated by word of mouth, the news media, and the existing extension system. In nearly all countries, plans for sowing semidwarf varieties were overfulfilled. In fact, demand was so strong that seed prices skyrocketed (8).

Before turning to individual country efforts and performance, it might be well to establish perspective by briefly discussing the history of semidwarf wheat production in Mexico, the originating country. The high-yielding semidwarf wheat varieties were released and quickly adopted by Mexico's growers in the early 1960's. They now account for over 90 percent of Mexico's total wheat area (46). Wheat yields increased from an average of 1,450 kilograms per hectare in 1957-59 to 2,670 kilograms in 1967-69. Current reports indicate that the national yield average is now 3,000 kilograms per hectare. As a result of higher levels of production, Mexico has become self-sufficient in wheat. Despite a considerable reduction in the area sown to HYV wheat, the Government has even had to cope with surplus wheat supplies (6, 7, 8, 14).

Mexican research work has been very active. CIMMYT and the National Institute of Agriculture Research (INIA) work very closely in all types and phases of researches at both local and international levels. (See section on CIMMYT, below).

TECHNOLOGICAL EFFECTS OF HIGH-YIELDING VARIETIES

The higher levels of 1968-70 wheat output in India and Pakistan associated with semidwarf technology are a sharp contrast to historical levels and trends. Figure 3 shows production in these years, compared with the linear trend for 1950-67. The fact that the 1968-70 observations fall far above the trend line is indicative that there has been a radical change in some aspect of wheat production from the previous pattern of development. Extrapolated 1950-67 trends would not reach the current levels of output in India and Pakistan until the end of the century.

Conceptually, semidwarf technology has caused the production function for wheat to shift upward. 8/ This is illustrated in figure 4 with supply curve S_1 representing the production function for the local wheat varieties and supply curve S_2 for the semidwarf varieties. At a given level of inputs per unit of land (I_1), there would be a yield of Q_1 units from S_1 and Q_2 units from S_2. The latter would exceed the former by Q_2-Q_1. The greater responsiveness of the semidwarf varieties to input changes is also illustrated in figure 4. If the level of inputs is increased from I_1 to I_2, the corresponding yield increases are Q_3-Q_1 for the local varieties (S_1) and Q_4-Q_2 for the semidwarfs. It is clear by visual inspection that Q_4-Q_2 is substantially greater than Q_3-Q_1.

8/ See (37) for a fuller discussion of this concept.

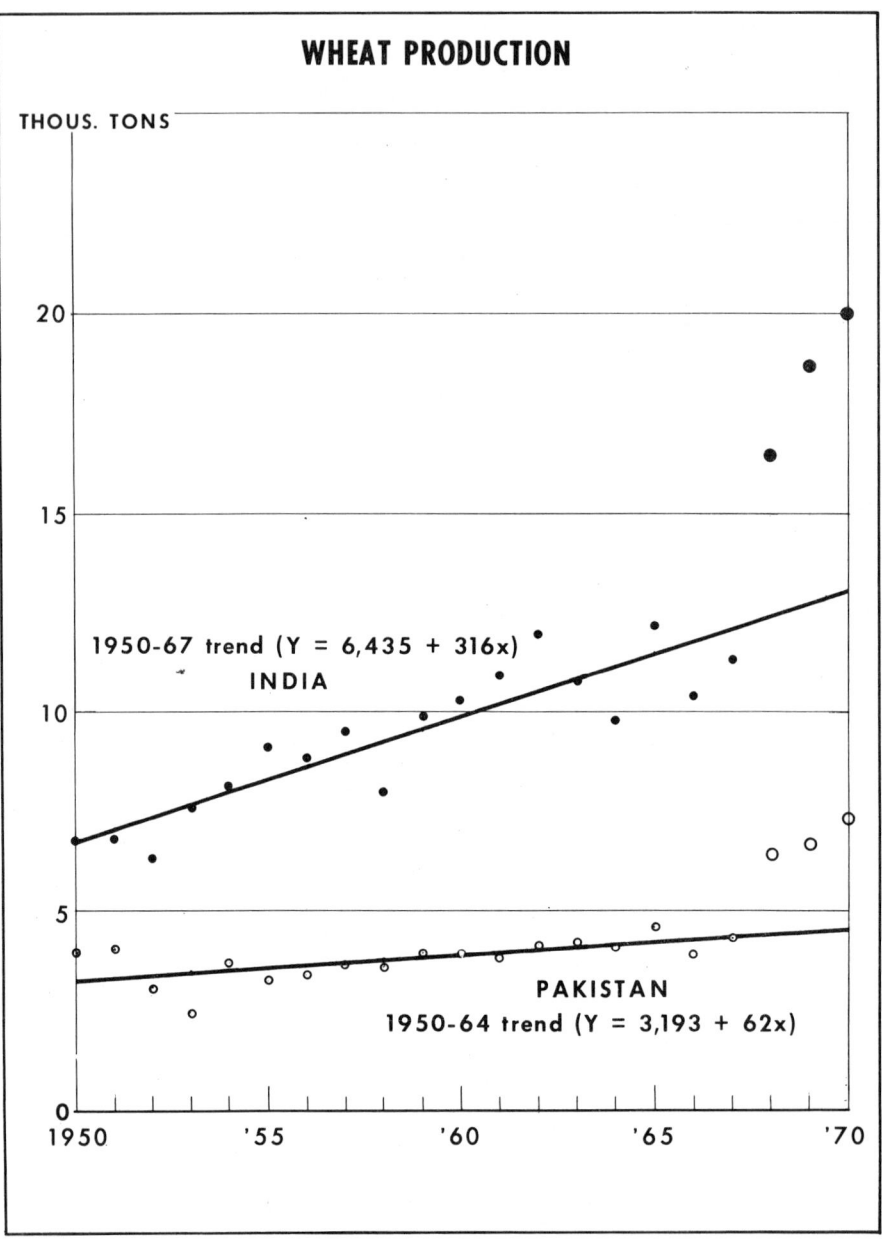

HYPOTHETICAL PRODUCTION FUNCTIONS FOR LOCAL AND HIGH-YEILDING VARIETIES OF WHEAT

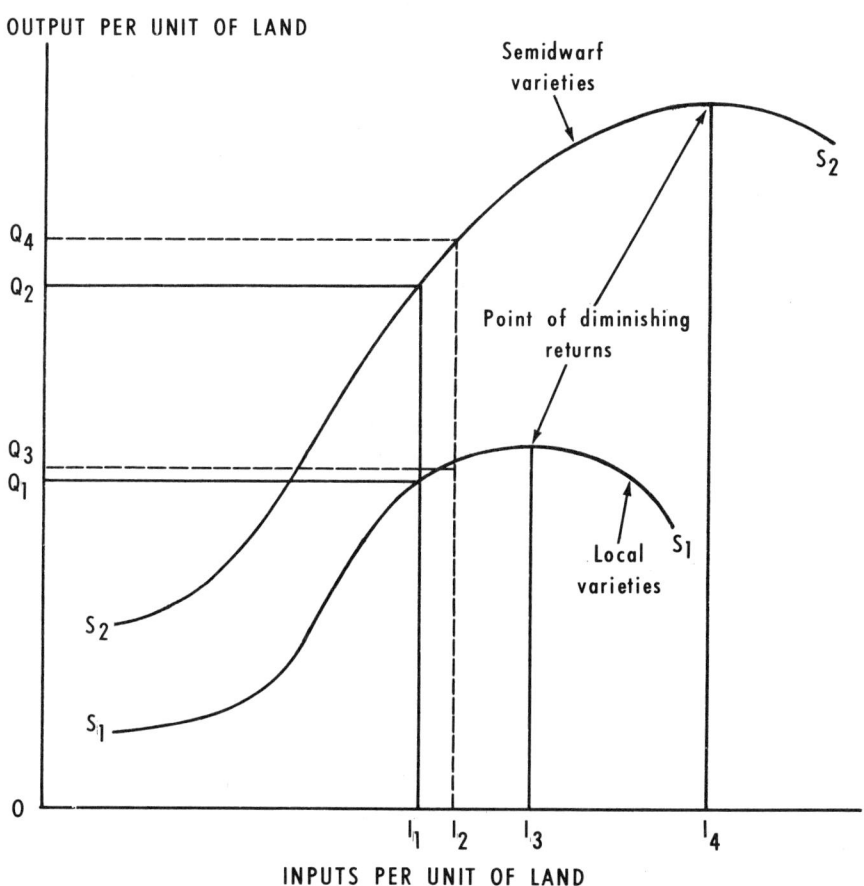

Figure 3 also demonstrates that at higher input levels, diminishing returns would set in for local and semidwarf varieties at input levels of I_3 and I_4. However, this occurs sooner for the local varieties (I_3). Again, this phenomenon reflects the greater input absorptive capacity of the semidwarf varieties.

The graphic evidence above supports the general recognition of a major technological breakthrough in wheat production. It is clear that future expectations should not be cast on the trends of the past, and that the stage has been set for a new growth pattern.

COUNTRY SITUATIONS

The following sections focus on the development, current situation, and trends of semidwarf wheats in the countries where they have been introduced. Where information is available, attention is given to factors limiting the expansion of semidwarf outputs as well as those contributing to expansion. These include land, climate, water availabilities, input supplies, agricultural institutions, and farm policies.

INDIA

Development

The first major shipment of 250 tons of semidwarf wheat from Mexico was imported by India in 1965. It was composed of the Mexican varieties, Lerma Rojo and Sonora 64, and was sown on about 3,000 hectares with excellent results. Thus, when the food situation in India became aggravated by the droughts in 1965 and 1966, the Government moved swiftly to import 18,000 tons of semidwarf wheat seeds from Mexico. This was the largest shipment of its kind in Indian agricultural history. The administrative speed and efficiency with which it was handled made the transaction even more remarkable. The seed was distributed in time for sowing the 1967 crop. Concurrently, the Government of India strengthened the package program 9/ for farmers by suggesting improved cultural techniques and providing necessary inputs (6).

India's first major harvest of semidwarf wheat in 1967 was very successful despite relatively poor weather. An area of 0.5 million hectares, nearly all of which was irrigated, produced 1.3 million tons of semidwarf wheat; yields averaged 2,500 kilograms per hectare as opposed to only 820 kilograms per hectare for local varieties (table 3). This success constituted the first stage of India's shifting production function for wheat. Using seed from the 1967 crop and seed released from India's own breeding program, the semidwarf wheat area in 1968 was boosted to 3 million hectares, which accounted for 20 percent of the total. With very favorable growing conditions and increased application

_____
9/ See footnote 6/.

Table 3.--India: Area, production, and yield of wheat; local and high-yielding varieties, annual, 1966-70

Items	Year harvested				
	1966	1967	1968	1969	1970
Area	- - - - - - - - - 1,000 hectares - - - - - - - - -				
Local.	12,653	12,324	12,056	11,165	10,515
Semidwarf	3	514	2,942	4,793	6,111
Total.	12,656	12,838	14,998	15,958	16,626
Production	- - - - - - - - - 1,000 tons - - - - - - - - - -				
Local.	10,416	10,115	8,698	7,460	6,487
Semidwarf.	8	1,278	7,843	11,191	13,606
Total.	10,424	11,393	16,540	18,651	20,093
Yield	- - - - - - - - - - - Kg./ha. - - - - - - - - - - -				
Local.	823	821	721	668	617
Semidwarf.	2,667	2,486	2,666	2,335	2,226
Average.	824	887	1,103	1,169	1,209

Sources: Area and production same as table 2; yields computed from data in table 2.

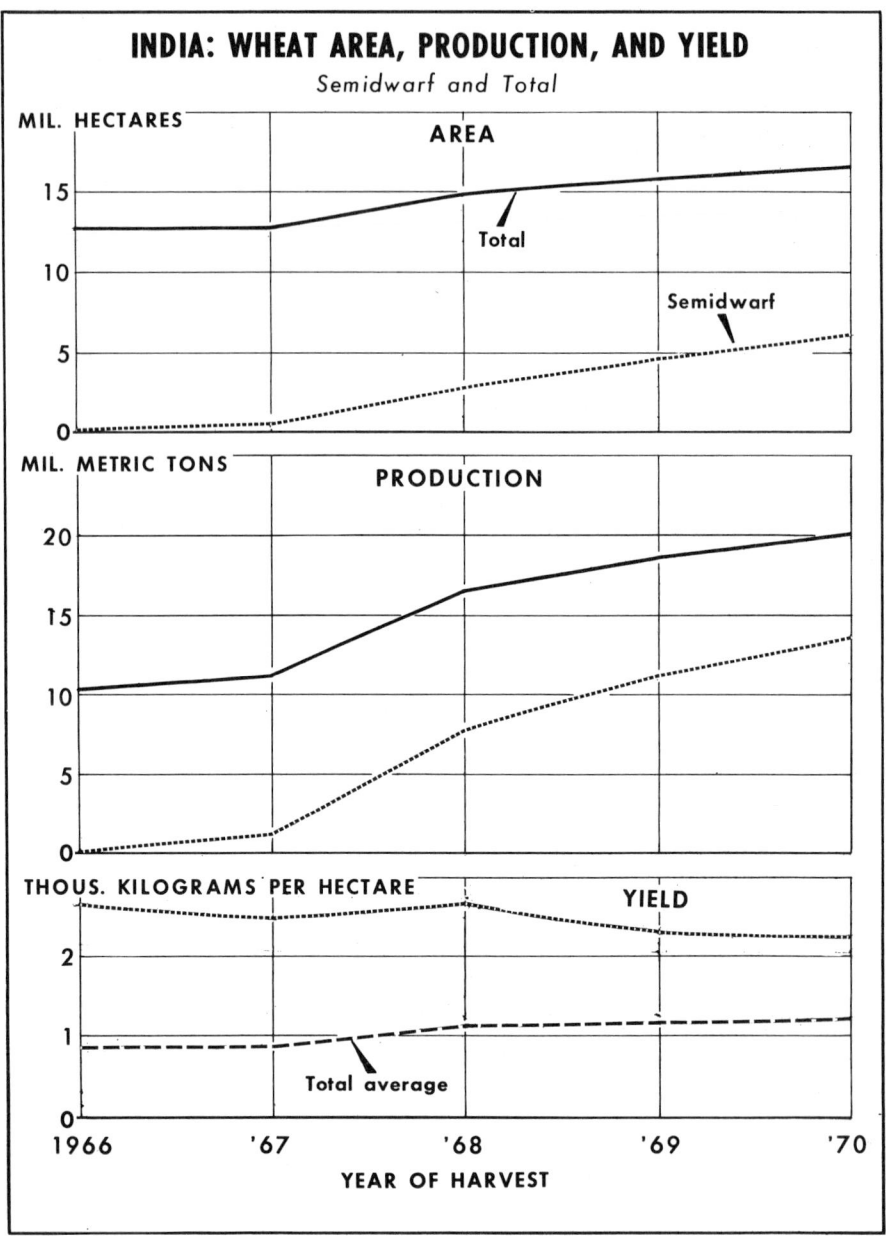

of fertilizer, 1968 production of semidwarf wheat soared to 7.8 million tons and accounted for nearly one-half of the then record wheat harvest of 16.5 million tons. Average semidwarf yields, which increased to 2,700 kilograms per hectare, were particularly impressive in view of the greatly expanded sown area. The 1968 crop eliminated any lingering doubts about the potential of the semidwarf varieties and substantiated the view that a technological breakthrough had occurred and had been successfully implemented in the field (7). The semidwarf area in 1969 jumped to 4.8 million hectares, accounting for 30 percent of the total wheat area. Semidwarf yields dropped 12 percent however, because weather conditions were decidedly less favorable. But the yield levels were still relatively high and semidwarf output, which reached 11.2 million tons, accounted for 60 percent of the record crop of 18.7 million tons (8).

Situation and Outlook

India's third consecutive record wheat crop was harvested in 1970. Weather conditions were generally near normal though there was a delay of 5 to 8 weeks in the winter rains which affected the northern wheat-growing areas. The semidwarf area increased substantially to 6.1 million hectares. 10/ Yields still remained at a reasonably high level. Production increased 22 percent to 13.6 million tons, amounting to nearly 70 percent India's total wheat crop. Thus, in 1969 and 1970, the output of semidwarf wheat varieties exceeded that of indigenous varieties.

Initial weather conditions for the 1971 crop were very favorable. The June-September 1970 monsoon was the best since 1967. Soil moisture was sufficient for proper seeding, germination, and early growth. However, the October-December period was dry in the northern and central areas and may have retarded normal growth. But there have been no reports of unusual crop damage, owing in part to the large irrigated area for wheat. Winter rains commenced in late December and were expected to boost the 1971 crop (89). With increased application of inputs and expanded use of high-yielding varieties the outlook for India's 1971 wheat crop appears good.

Research Activities

The All-Indian Wheat Improvement Program was first concerned with the multiplication of imported semidwarf wheat to meet the expanding demand for seed. Since then, it has successfully conducted a program to cross local varieties with the imported Mexican semidwarf varieties. The imported varieties, while high yielding, were soft spring wheats with a relatively low protein content. In addition, the dark red color of the kernel was disapproved of by consumers who preferred the amber color of the local wheats for making chapattis 11/. There are 10 major research centers, 8 secondary centers, and many more research stations throughout the country in the program. The major ones are located at New Delhi, Ludhiana, and Pantnagar. Some of the resulting crosses retaining the high-yielding characteristic, have proven more suitable to consumers taste and preference. In 1967, several amber-colored highly rust-

---

10/ An unofficial early estimate was about 5.1 million hectares (12).
11/ Chapati--a pancake-like, unleavened bread, which is a basic food in India and West Pakistan (24).

resistant selections were released for commercial production under the names of S-227, Kalyansona (two-gene dwarf wheat), Sonalika, Safed Lerma, and S-331. An amber-seeded, high protein strain developed by treating Sonora 64 with gamma rays was also released under the name of Sharbati Sonora (7, 8).

On May 30, 1970, India released the first Triple Dwarf (3-gene) wheat variety--UP 301. (9). According to Dr. J. P. Srivastava of the Agricultural University of Uttar Pradesh, UP 301 possesses a high degree of resistance to most rusts and gives a fairly satisfactory yield. Moreover, its protein content is one-third higher than that of Kalyansona. Its 71-76 centimeter (28-30 inches) height, which is shorter than the original Mexican varieties makes it even more resistant to lodging. Its most attractive characteristic is a strong gluten content which is highly desirable for chapattis.

PAKISTAN 12/

Development

Semidwarf wheat varieites were introduced in Pakistan as early as 1960 by agricultural scientists who returned from a training program in Mexico with a small sample of seeds. Tests from this sample indicated that these high-yield varieties were adaptable to growing conditions in Pakistan. In mid-1964, Pakistan embarked on its Accelerated Wheat Improvement and Production Program, sponsored by the Government with the financial and technical assistance of CIMMYT and the Ford Foundation. The Government organized its economic and technical forces to achieve the designed objectives. In 1965, 350 tons of Penjamo 62 and Lerma Rojo 64 semidwarf seed were purchased from Mexico and sown on about 5,000 hectares (table 4). Yields averaged 2,800 kilograms per hectare--more than triple the average of local varieties. The severe drought in 1965-66 prompted officials to take bold actions to alleviate a mounting food crisis. In 1967, a purchase of 42,000 tons of semidwarf seed was made in Mexico and distributed to farmers in time for fall sowing. This represented the largest single purchase of imported wheat seeds ever negotiated. Nearly one million hectares or 16 percent of the total wheat area were seeded with these varieties and the resulting 1968 harvest was 2.2 million tons or one-third of the total wheat crop. Even though this was the first time that semidwarfs were grown on a large scale, yields averaged 2,300 kilograms per hectare.

The 1968 crop was the basis for a large seed stock with which the semidwarf wheat area was expanded to 2.4 million hectares the following year. Semidwarf yields averaged 1,700 kilograms per hectare in 1969 and output accounted for three-fifths of total wheat production. The substantial decrease in yields was attributed to the expanded area (more than double) and less favorable weather. Yet HYV yields were still well above local varieties.

12/ The discussion in this section refers to West Pakistan, where nearly all of the wheat is raised.

Table 4.--Pakistan: Area, production, and yield of wheat; local and high-yielding varieties, annual, 1966-70

Items	Year harvested				
	1966	1967	1968	1969	1970
Area	- - - - - - - - 1,000 hectares - - - - - - - - -				
Local.	5,205	5,316	5,104	3,657	3,516
Semidwarf.	5	101	957	2,388	2,833
Total.	5,210	5,417	6,061	6,045	6,349
Production	- - - - - - - - - 1,000 tons - - - - - - - - - -				
Local.	3,938	4,170	4,240	2,762	3,049
Semidwarf.	14	224	2,237	3,949	4,350
Total.	3,952	4,394	6,477	6,711	7,399
Yield	- - - - - - - - - - - Kg./ha.- - - - - - - - - - -				
Local.	757	784	831	755	867
Semidwarf.	2,800	2,218	2,338	1,654	1,535
Average.	759	811	1,069	1,110	1,165

Sources: Area and production same as table 2; yields computed from data in table 2.

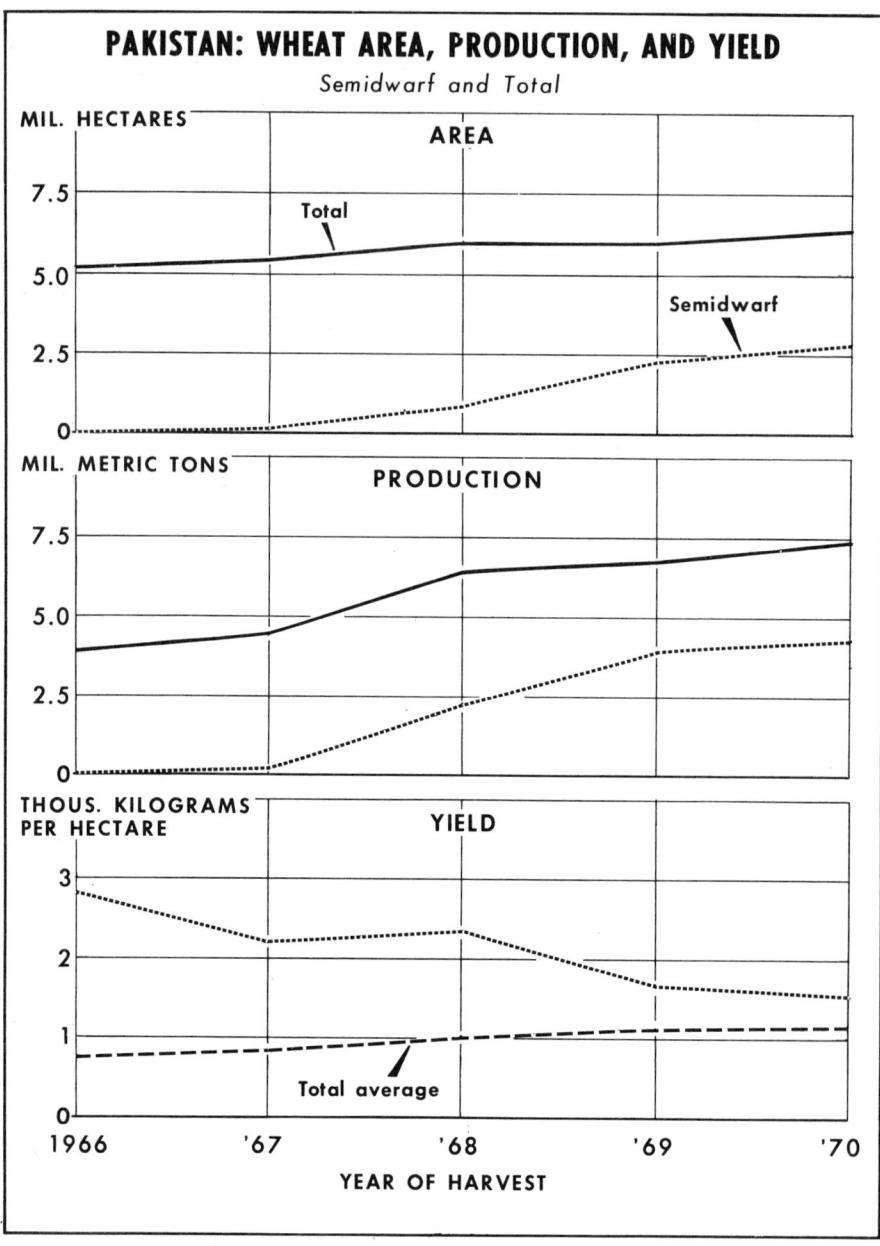

Situation and Outlook

On the whole, weather conditions in West Pakistan during the 1969-70 wheat growing season were satisfactory. However, in barani (rainfed) areas, drought occurred during November and December 1969 and the early part of 1970. Late January rains improved the situation markedly. If there was some adverse effect, semidwarf wheat would have suffered the least because of irrigation. Estimates relative to 1970 wheat output released by the Ministry of Agriculture and Works show an all time record of 7.4 million tons, an increase of 11 percent over the previous year. Total area also increased 5 percent over 1968-69, mostly for the high-yielding varieties. There was an estimated increase of 10 percent in semidwarf output accounting for close to 60 percent of total wheat production (table 2).

Lack of moisture during the 1970-71 sowing period resulted in a lower nonirrigated wheat area. Under normal conditions, there is usually adequate rainfall in November and December for germination and growth. However, if general drought conditions prevail, it might affect the total wheat output somewhat. It is estimated that the 1970-71 area could be at least the same or slightly above that of 1969-70. However, the area shifts from local to high-yielding varieties would be substantial, possibly up more than 20 percent over the previous year.

Research Activities

The Accelerated Wheat Improvement and Production Program, encouraged by the Government of Pakistan, has sponsored an intensive wheat research operation to increase the yields of semidwarf varieties and to improve semidwarf quality in terms of nutrition and consumer acceptance. The original varieties imported from Mexico have been crossed with indigenous varieties. Mexipak 65 is one such crossed progeny that is now being grown extensively in Pakistan. It not only has the high-yielding quality of the Mexican varieties but it is amber in color, thus increasing consumer acceptability for baking chapattis.

In addition, widespread microtrials in the past have established the superiority of three new varieties: Mangla 68, Norteno 67, and Iria 66. The first two are white grains and the third a red-type grain. They have been widespread in cultivation and have strengthened resistance to rust. Current activities concentrate on the yield performance, adaptability, and disease resistance of several promising lines. One such line in the evaluation stage is the triple-gene variety selected from the Mexican Cross II-23584.

The breeding program in Pakistan is broad and dynamic and is supported by CIMMYT with superior genetic materials. It also has extended its activities to agronomic and soils research recently.

TURKEY

Development

Turkey became one of the pioneers in raising semidwarf wheat when the Turkish Agricultural Research Station at Adapazari obtained 17 sample breeding stocks from Mexico in 1959 for testing. In 1965, 40 kilograms of semidwarf seed of Sonora 64 and Lerma Rojo were obtained by USAID from India. They were sown on a private farm in Tarsus and the resulting yields were surprisingly high. The excellent performance and exceedingly high yield of the semidwarfs aroused interest among neighboring farmers. As a result, a group of about 100 wheat cultivators in the Cukurova area pooled their financial resources and with Government assistance, procured 60 tons of Sonora 64 from Mexico. This lot was sown during the fall of 1966 in Adana region (23).

Weather was extremely favorable for grain production during the 1966-67 crop year and the 1967 wheat harvest totaled 9.0 million tons, a record high for Turkey (table 5). Naturally, the semidwarfs benefited from the favorable weather. Yields for the high-yielding varieties averaged 3,300 kilograms per hectare and on some experimental plots ran as high as 6,620 kilograms. In view of these encouraging facts, the Government decided to expand the semidwarf wheat area and established a wheat "crash program" for 1967-68.

At the direction of the Minister of Agriculture, Turkey purchased 22,000 tons 13/ of semidwarf wheat varieties, chiefly Penjamo 62, Lerma Rojo 64, Super X, and Mayo 64, from Mexico. Unfortunately, 1967-68 weather conditions were unfavorable. At the start, fall sowing was delayed by harvesting problems with cotton and sugar beets, which precede wheat in the multicropping system. Excessive rains slowed the cotton harvest. Earthquakes, which damaged sugar factories, also led to a delay in the beet harvest. As a result, only 17,000 tons of the available supply of semidwarf seed were sown, leaving a shortfall of 5,400 tons. The area sown to semidwarf varieties which totaled about 170,000 hectares, was centered in the lower plains in the Marmara, Aegean, and Mediterranean coastal regions. Following the sowing difficulties, there was a flood in the Adama region in January 1968, a cool spring, and then a dry spell in late May which adversely affected production.

Because of these unfavorable natural conditions, total wheat output for 1968 was down 600,000 tons from the previous year's record. Areas around the Anatolian Plateau suffered the most. 14/ Despite these adversities, the semidwarf varieties performed very well. They accounted for only 2.3 percent of the total area but yielded 7.1 percent of the total wheat crop. The HYV yields averaged 3,500 kilograms per hectare, a level indicative of success.

With the confidence of successful experience, cultivators and the Government, together with the assistance of international agencies, formed a "second

---

13/ 400 additional tons of semidwarfs were purchased from the United States (23).
14/ The Anatolian Plateau produces about 65-70 percent of Turkey's wheat.

Table 5.--Turkey: Area, production, and yield of wheat; local and high-yielding varieties, annual, 1966-70

Items	Year harvested				
	1966	1967	1968	1969	1970
Area	1,000 hectares				
Local.	7,163	7,203.4	7,134	7,071	6,937
Semidwarf.	n.a.	0.6	170	579	623
Total.	7,163	7,204	7,304	7,650	7,560
Production	1,000 tons				
Local	8,200	8,998	7,805	6,672	6,203
Semidwarf.	n.a.	2.0	595	1,628	1,797
Total.	8,200	9,000	8,400	8,300	8,000
Yield	Kg./ha.				
Local	1,145	1,249	1,094	944	894
Semidwarf.	n.a.	3,279	3,500	2,812	2,884
Average.	1,145	1,249	1,150	1,085	1,058

Sources: Area and production same as table 2; yield computated from data in table 2.

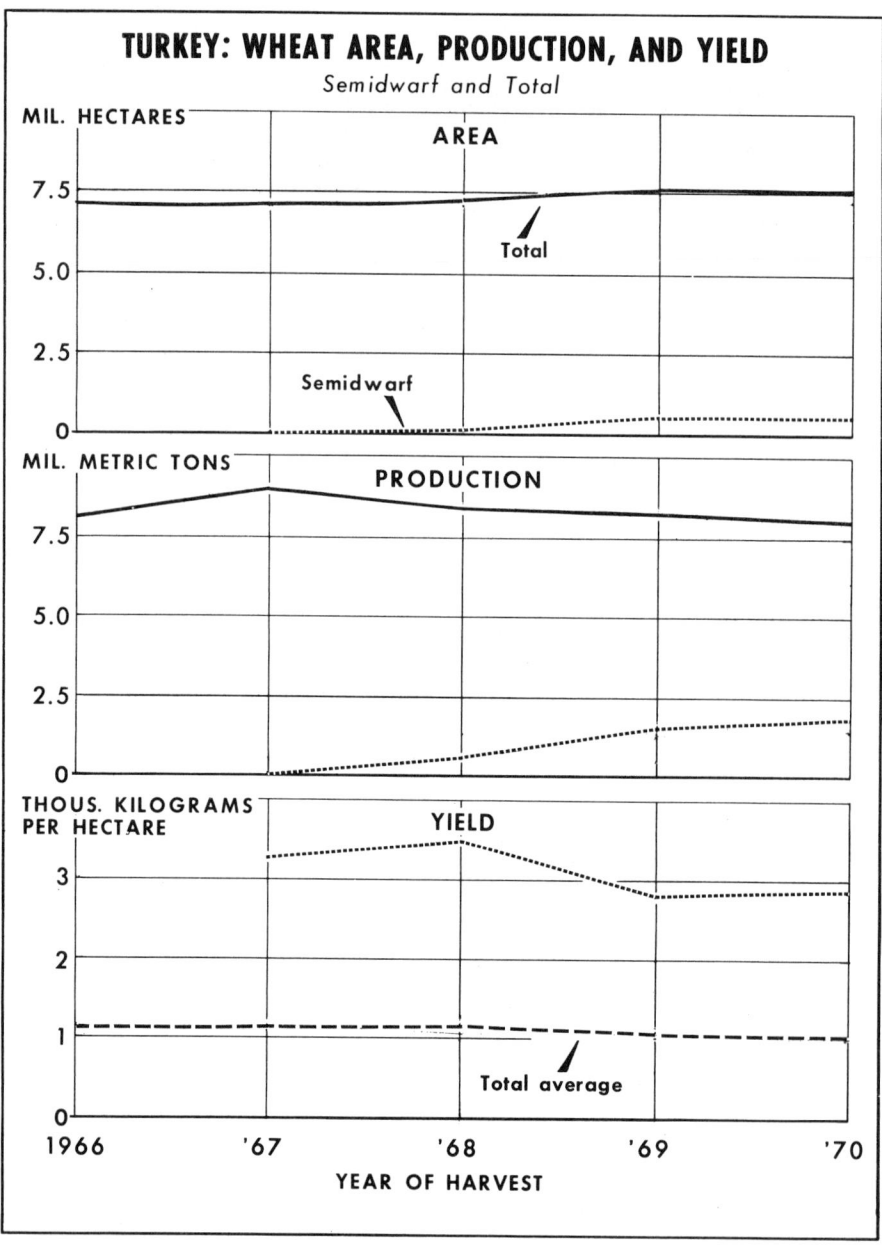

crash wheat program" for 1968-69. The multiplication of semidwarf seeds by Government farms plus the stock carried over by cultivators could not meet the huge demand. Some grain dealers were reported to have purchased early harvested semidwarf wheat at relatively low prices and shipped it to other areas for seed at a higher price. Of the 67 provinces, 24 were brought under semidwarf wheat cultivation. They covered the coastal areas where irrigation was readily available, and some interior plateau provinces. There was considerable substitution of semidwarf wheat for cotton, pastures, and other crops (23).

Weather conditions were very favorable during the late fall of 1968 and early spring of 1969. However, late spring rains and subnormal temperatures adversely affected the crop. There was also some winter kill on the Anatolian Plateau and floods in the Mersin area early in January.

The total 1969 wheat area was about 5 percent larger than the previous year; but production was down slightly. Under the second crash program, the semidwarf area expanded from 170,000 to 650,000 hectares. Semidwarf output surged from 600,000 to over 1.6 million tons. Though the yield of high-yielding varieties was down from previous year's high level, it was still substantially higher than that of local varieties. The unfavorable climatic conditions admittedly weakened yield potentials, but insufficient supplies of fertilizers and other inputs as well as less supervision of the cultivators under the crash program were responsible for the lower yield. Yet, had it not been for the semidwarf program, Turkey's 1969 total wheat harvest would have declined substantially.

Situation and Outlook

Total wheat output declined in 1970, marking the third consecutive disappointing crop. Adverse weather was again responsible. Late fall rains delayed sowing and retarded germination. Later, temperatures hovered around freezing in the central Anatolian and Thracean Plateaus, causing winter kill. The poor growing season was climaxed by the driest May and June on record. But most of HYV wheat grown in the coastal regions performed reasonably well. A harvest of 1.8 million tons accounted for 22 percent of the total output of 8.0 million tons. The most encouraging fact was that semidwarf yield was up 10 percent from 1969.

Although the gain of HYV wheat did not fully offset the reduction of the local varieties, it should be realized that if the area given to semidwarf wheat in 1970 had been sown with local varieties, total wheat production would have been at least 1.0 million tons lower. 15/

Thus far, weather conditions appear favorable for the 1970-71 wheat crop year, suggesting a departure from the previous 3 year years' unpleasant experiences. Timely fall rains encouraged cultivators to sow earlier, particularly

---

15/ This estimate is based on a yield of 1086 kilos per hectare (the 1961-65 average increased by 5 percent to take into account trend) and the semidwarf area of 650,000 hectares. This approach could be applied to the earlier years as well.

in the Anatolian Plateau areas. Germination had been fairly good, and wheat growing conditions are normal. The semidwarf wheat area is reported to have expanded substantially, compared with 1969 and 1970. If favorable climatic conditions continue and spring rains, particularly in May, are adequate, it is anticipated that the 1971 harvest will be better than that of recent years (90).

Research Activities

Since the early introduction of the HYV, the Government has launched a varietal improvement research program. In collaboration with the Rockefeller Foundation and CIMMYT, a Cooperative Center of Research and Training in Wheat Improvement was established in 1969 for Turkey and neighboring countries. A primary objective of the research program is to improve the semidwarf varieties with spring habits for the coastal regions and to develop high-yielding varieties with winter or semiwinter habits 16/ for the higher elevations of the Anatolian Plateau. A durum wheat-breeding program is also underway to develop frost-resistant varieties from Mexican wheat for higher-latitude areas. The research program of the Center also includes aspects of entomology and plant pathology (8).

AFGHANISTAN

With the help of USAID and CIMMYT, Afghanistan has developed a successful wheat program. With an initial shipment from Pakistan of 190 tons of seed, about 1,800 hectares were sown to semidwarf wheat in 1967 (table 2). The area was expanded to 22,000 hectares in 1968, and 122,000 in 1969. An estimated 150,000 hectares were sown to semidwarf wheat in 1970, accounting for 5 percent of the total wheat area. The HYV's yields have been good and these varieties now account for 14 percent of the total wheat output. However, the fact that much of Afghanistan's wheat is in high-elevation areas, which are cold and dry, is a constraint to HYV expansion (8). 17/

TUNISIA

Tunisia cultivates about 950,000 hectares of wheat, virtually all of which is rainfed crop. Average annual production during 1960-69 was about 410,000 tons and the yield was 430 kilos per hectare. Output only fills 50 percent of Tunisia's requirements. Approximately, two-thirds of the production is bread wheat, the remainder is durum.

With assistance from international institutions and agencies, Tunisia in 1967 launched The Project for Increased Cereal Production (ACPP). Five

---

16/ These are derivatives from crosses between spring and winter varieties developed by Dr. J.A. Rupert at Davis, California.
17/ The reader is reminded that the semidwarf varieties developed in Mexico are spring wheats which do not have the dormancy characteristic needed to withstand cold weather.

agronomists from CIMMYT were assigned to help implement this program. Three Mexican dwarf varieties (Inia 66, Tobati 66, Sonora 63) were sown on 800 hectares. The 1967-68 crop was very successful; 1,500 tons of HYV wheat were harvested and yields far exceeded local varieties. In 1969, the HYV area expanded to 13,000 hectares. With favorable weather, the national wheat program was spectacularly impressive during the 1968-69 season. The HYV production was about 24,000 tons and yields reached as high as 3,000-4,000 kilograms per hectare. The achievements encouraged the Government to boost the HYV area to 53,000 hectares for the 1969-70 crop. With more experienced cultural practices and normal weather conditions, HYV output in 1970 increased fivefold over 1969 to 120,000 tons. However, local varieties in the south and central areas were adversely affected by weather and outbreaks of rust-damaged durum.

A goal of 100,000 hectares was set for the 1971 HYV area. With favorable weather, this area could produce as much as 150,000 tons of HYV wheat. However, rains during late 1970 delayed sowing, possibly causing below average yields.

The varietal improvement section of the Wheat Project in ACPP was established in cooperation with INRAT (National Institute of Agricultural Research of Tunisia) and the assistance of CIMMYT and USAID. It covers a very broad research area. The excellent start during 1968-69 and the enthusiasm already generated by the success of the production program, may well evolve into a base for general improvement throughout North Africa (6, 7, 8).

IRAN

Iran has a wheat area of nearly 4 million hectares (1960-69 average). The long-time national average yields are low, about 840 kilograms per hectare; with 600 kilograms per hectare on rainfed land and 1,200 kilograms per hectare on irrigated land. To improve yields, Iran started wheat breeding over 30 years ago. Technical assistance has been obtained from international agencies and institutions. Varietal improvement is pursued throughout the wheat area; however, the major wheat problem is rust disease.

In 1969, Iran first sowed the high-yielding varieties of wheat on 10,000 hectares. The seed stock (Penjamo 62, Inia 66) originated from Mexico, but was imported from Turkey. The HYV harvest was 16,000 tons in 1969 and the yields substantially exceeded the native varieties. Although rusts occurred and some crops were damaged, grains of HYV (particularly Inia 66) filled out well in all locations. Encouraged by this, Iran expanded its HYV area to 90,000 hectares in the 1969-70 crop season by using 1,500 tons of seed stock (Penjamo 62) imported from Denmark as well as seed retained from previous semidwarf crops. The substantial increase of semidwarf area boosted 1970 output to 120,000 tons, although total wheat harvest decreased from 3.9 million tons in 1969 to 3.8 in 1970. This decrease was due to the shortage of rainfall.

The Government plans to increase the wheat area for 1970-71 by 70,000 hectares. However, because of prolonged dryness in the wheat area in 1969-70, the size of the 1970-71 crop remains to be seen.

Iran's northern plains are considered promising for ground water development. Many local varieties have already been replaced in much of the area. Government policy to encourage wheat production encounters increased competition from other crops for the same fertile soils. However, there is still a good deal of uncultivated area in mountain regions where rain is adequate. These areas should eventually be exploited for the high-yielding varieties and new technology.

The present HYV collection and distribution program could be made more efficient and the results improved. Agronomic research should be conducted immediately to determine what modifications in cultural practices are needed to further adapt the dwarfs to Iranian conditions (8).

MOROCCO

Morocco began its semidwarf wheat production in 1968 with an area of 200 hectares of Siete Cerros, Inia 66, Penjamo 62, and Tobari 66. Owing to the good yields realized, 500 tons of five semidwarf varieties were imported the next year from Mexico and planted on an area of 5,000 hectares. Unfortunately, excessive rainfall caused a very serious epidemic of septoria leaf and glume blotch on both semidwarf and local varieties. Still, the HYV greatly outperformed indigenous varieties.

The HYV area expanded considerably in 1970 to 10,000 hectares. The 1970 semidwarf wheat harvest is estimated to be 20,000 tons against only 7,000 tons in 1969; the increase is attributed to expanded area, favorable weather, and fewer disease problems.

The 1971 outlook appeared favorable after rains in late December 1970 broke a dry spell. However, output still depends upon weather conditions during the remainder of the crop year (table 2).

Although Morocco possesses extensive lands suitable for wheat, erratic and unseasonable rainfall is a negative factor in expanding the semidwarf wheat area. The HYV's have performed well under adequate rainfall conditions (70, 71, 72).

A cooperative project with CIMMYT to find disease-resistant varieties is now in progress. More than 2,000 lines and varieties of experimental wheats, mostly of Mexican origin, were grown.

OTHER COUNTRIES

Other countries, though not yet adopting semidwarf wheat on a significant scale, have been preparing for future introduction. Countries in this group are engaged in plot testing and varietal improvement for disease resistance, higher yields, and better quality.

Wheat production in Nepal accounts for only 3 percent of the total food grain output (rice 68 percent, corn 25 percent, and millet 4 percent). Since Nepal is very mountainous, the growing area is centered in Katmandu Valley and Terai, an extension of the Gangetic Plain. In recent years, with the assistance of USAID, the Ford Foundation, India (Indian Aid Program), and other agencies, Nepal has been active in growing high-yielding cereal varieteis, particularly semidwarf wheat. In 1965, Nepal imported 38 tons of Mexican wheats (mainly Lerma Rojo) from India, from which a good harvest was realized. In the following year, 450 more tons of the same variety were shipped from India. Reports indicate that despite the unfavorable weather (mostly droughts) since 1964, Nepal has continued the wheat expansion program. Area and production data pertaining to HYV and total wheat are very sketchy. However, some sources estimated that the HYV area has increased from 1,400 hectares in 1966 to 75,000 in 1970. It may be assumed that Nepal's HYV wheat program has contributed to its food supply. This assumption is evidenced by the continuous area expansion and extensive assistance rendered by outside agencies.

In cooperation with CIMMYT, Argentina has been concentrating on research and breeding since 1968. Ninety advanced generation lines of the semidwarf wheat were included in the yield tests. Argentina's breeding research includes durum wheat (8).

With the support of the Ford Foundation, Lebanon has been conducting the "Arid Lands Agricultural Program." A cooperative wheat improvement program for a number of Near and Middle East countries, including Jordan, Iraq, and Iran, has also been developed. The program is closely related to CIMMYT and Ford Foundation activities. This spring wheat breeding program is also being delivered for the coastal areas of Turkey (18, 19). An intensive research program for semidwarf varieties in North Africa is centered in Tunisia. All these research works are carried out by the Project for Increased Cereal Production (ACPP) and the National Institute of Agricultural Research of Tunisia (INRAT) with the cooperation of CIMMYT, USAID, and other agencies. Attacks of powdery mildew were heavy throughout the regions, permitting selection for resistant lines. All known insects of North Africa are being used in tests to improve resistance (8).

Colombia initiated a sound and successful wheat program relatively early. The Rockefeller Foundation has provided extensive technical assistance. Later the Colombian Agricultural Institute (ICA) was established and has since developed more than a dozen improved wheat varieties. The impact of this program has practically reached all farmers. However, wheat is a minor crop in Colombia (about 3 percent of total crop area); cultivators and the Government tend to shift their interest from wheat to other more profitable crops. Barley, for example, the major crop for the beer industry, is in great demand. Consequently, the wheat program so far has been flagging. Only 8 percent of Brazil's food crop is wheat; its major food staples are rice and corn. With the assistance of FAO, Ford Foundation, CIMMYT, and USAID, wheat production increased from a low of 120,000 tons in 1963 to 360,000 in 1967, 600,000 in 1968 and 1,730,000 in 1970. Improved seeds have been used widely on the growing area centers on Rio Grande do Sul in the South. The relatively high price of wheat against rice has stimulated a significant shift in production from rice and other crops to wheat. The National Wheat Program of Paraguay began in 1965, and the improved seeds have been used since then. USAID has been providing technical assistance rather actively in this country.

CIMMYT

Research on new wheat varieties has been intensified since the breakthrough of the semidwarfs. Under the leadership of CIMMYT, coordination among many countries throughout the world has been expanded and strengthened. Major research projects include the blending of germ plasm complexes of spring wheat and hard winter wheat, in the hope of developing new winter, semiwinter, and spring varieties with higher yields. Hybrid wheat seed is also under intensive research. CIMMYT has developed restorer lines capable of restoring fertility to certain cytoplasmic sterile male types of wheat. Another project involves crossing Mexican semidwarf wheat with rye to develop triticales which may even outyield the HYV's. The improvement of wheat quality--higher protein and essential amino acid content--has gained substantial ground. Research work on durum has developed an excellent program. Many are light insensitive, have fair fertility, and are disease resistant. Durum is also used as a parent in the triticale-breeding program. CIMMYT also has broadened the spectrum and increased the depth of the genes for resistance to rusts and other diseases. Research on three-gene wheat varieties has gained remarkable momentum. Some countries have even distributed some triple gene seeds for trial commerical production. Numerous experimental materials have been distributed on a worldwide basis by CIMMYT for further research purposes. [18]/

CONCLUSIONS

Based on past developments and recent situations of the semidwarf wheat program in the developing countries, it appears likely that the momentum generated in the programs discussed above will continue. It is expected that under normal climatic conditions wheat output will continue to increase. The pace of development may vary because each of the countries is at a different stage of program implementation.

However, the pace of development will largely depend on:

(1) The direction of governmental policies and efforts. If food were in

---

[18]/ See table 1.

relatively adequate supply, attention and priorities might be directed to other problem areas to achieve a national objective of total economic growth and development. However, regardless of circumstances, the program itself or at least phases of the program should now be self-perpetuating;

(2) The efficiency of the institutions and infrastructure within the framework of the regulated program. Future programs should be directed towards strengthening and coordinating activities for future expansion. Minor changes may occur from time to time to comply with the changing agricultural policies;

(3) The dissemination of knowledge to farmers. The package concept should be made known to cultivators. Apart from cultural methods and other requirements, the application of inputs to optimize output should be considered the key point for the whole operation. There is evidence in countries producing semidwarfs that cultivators' use of fertilizer is well below optimal levels for maximizing returns;

(4) The investment and performance of the private sector. Efforts and capital from the private sector should continue to flow into the programs; there is ample room for profits to encourage continuous investment. Greater involvement of the private segment would necessarily provide more permanency to the program. Hence, with a firm foundation, constant progress could be maintained without further major efforts by the Government;

(5) The area given over to semidwarf wheats. The area for raising HYV wheats could be expanded in areas currently used as well as in other parts of the world. However, at present, the greater portion would come from replacing those traditionally low-yield wheat-producing areas, where the irrigation system and water supplies can be improved. In some favorable ecological areas, much of the marginal lands can also be utilized profitably. The most hopeful approach for expansion is through intensive breeding and research. New varietal wheats possessing the capability of growing in areas beyond the traditional wheat belt would be the force for further expansion. Encouragement of multiple cropping would also benefit the semidwarf wheat program;

(6) The effectiveness of research on new improved varieties. Current research and the expanded research areas should provide the base for continuous varietal improvements. Public concern as to how soon and to what extent the next technological breakthrough will take place is unknown and unpredictable. The efforts, investment, and performance of the countries already involved in this program could serve as guidelines for evaluating the whole future program.

There may not be another surge in output such as that caused by the rapid and successful introduction of the semidwarf wheat in India and Pakistan. The success there occurred under very propitious conditions and situations. But with the wheat research activities now underway, technical improvements should continue to flow into the production programs at a relatively good rate. Therefore, we should be able to look forward to a continuing betterment of quality and yields in the countries under study in this report.

LITERATURE CITED

(1) Abbas, S.A.
1967. Supply and Demand of Selected Agricultural Products in Pakistan
1975. Oxford Univ. Press, Pakistan.

(2) Abel, Martin E. and Rojko, Anthony S.
1967. World Food Situation. U.S. Dept. Agr., Econ. Res. Serv., ERS-Foreign 35. August.

(3) Bell, David; Harrar, George; Myers, William and Hardin, Lowell
1968. Q and A on Agriculture, Science, and the Developing World. International Agriculture Research and Training Centers, New York, Oct.

(4) Borlaug, Norman E.
1968. Wheat Breeding and Its Impact on World Food. Third International Wheat Genetics Symposium, Australian Academy of Science, Canberra, Australia. Aug.

(5) Brown, Lester R.
1961. An Economic Analysis of Far Eastern Agriculture, U.S. Dept. Agr., Econ. Res. Serv., FAER-2. Nov.

(6) Centro International de Mejoramento de Maiz Y Trigo (CIMMYT) 1968. Report 1966-67 on Progress Toward Increasing Yield of Maize and Wheat. Mexico City, Mexico.

(7) _____
1969. Report 1967-68 on Progress Toward Increasing Yield of Maize and Wheat, Mexico City, Mexico.

(8) _____
1970. Report 1968-69 on Progress Toward Increasing Yield of Maize and Wheat, Mexico City, Mexico.

(9) _____
1970. UP 301, First Triple Dwarf Wheat Released in India. CIMMYT News, vol. 5, nos. 5-8, Mexico City, Mexico, May-Sept.

(10) Creupelandt, H. and Abbot, J.C.
1969. Stabilization of Internal Markets for Basic Grains: Implementation Experience in Developing Countries. Monthly Bulletin of Agricultural Economics and Statistics. Food and Agriculture Organization. Rome, Feb.

(11) Dalrymple, Dana
1969. Imports and Plantings of High-Yielding Varieties of Wheat and Rice in the Less Developed Nations. Foreign Agr. Serv., U.S. Dept. Agr. in cooperation with Agency for International Development, Apr.

(12) Dalrymple, Dana G.
    1971. Imports and Plantings of High-Yielding Varieties of Wheat and Rice, U.S. Dept. Agr., Foreign Econ. Dev. Serv., Foreign Econ. Dev. Rpt. 8., Jan.

(13) _____
    1969. Technological Changes in Agriculture. Effects and Implications for the Developing Nations. Foreign Agr. Serv., U.S. Dept. Agr. in cooperation with Agency for International Development, Apr.

(14) _____
    1969. Wheat and Corn in Mexico. Spring Review of the New Cereal Varieties. U.S. Dept. Agr., Mar.

(15) Direccion General de Estadistica
    1967. Anuario Estadistico Compendiado de los Estados Unidos Mexicanos.
    1966. Secretaria de Industria y Comercio, Mexico D.F.

(16) Dunbar, Ernest
    1968. India's Food Miracle. Look, vol. 32. no. 6, Mar.

(17) The Ford Foundation
    1967. A Richer Harvest, A Report on Ford Foundation Grants in Overseas Agriculture. New York City, N.Y., Oct.

(18) _____
    1969. The Ford Foundation Annual Report. New York City, N.Y., Oct.

(19) _____
    1970. The Ford Foundation, Annual Report, New York City, N.Y., Feb.

(20) French, J.T. and Lyman, P.
    1969. U.S. Agency for International Development. Spring Review of the New Cereal Varieties--Emerging Problems, Unpublished.

(21) Harrar, J.C.
    1950. Mexican Agricultural Program. Rockefeller Foundation. New York.

(22) Hendrix, William E., Naive, James J. and Adams, Warren.
    1968. Accelerating India's Food Grain Production 1967-69 to 1970-71. U.S. Dept. Agr. FAER-40. Mar.

(23) Humphrey, L.M.
    1969. Mexican Wheat Comes to Turkey, USAID/Turkey, Apr.

(24) Hutchison, John E., Naive, James J. and Tsu, Sheldon K.
    1970. World Demand Prospects for Wheat in 1980, U.S. Dept. Agr., Econ. Res. Serv., FAER-62. July.

(25) Hyslop, J.D. and Dahl, R.P.
    1968. Wheat Production in Tunisia, University of Minnesota, May.

(26) India, Government of
1970. Economic Survey 1969-70, New Delhi, India.

(27) _____, Ministry of Food and Agriculture
1968. Agricultural Statistics Report. India. Nov.

(28) _____, National Council of Applied Economic Research
1962. Long-Term Projections of Demand for and Supply of Selected Agricultural Commodities 1960-61 to 1975-76 (India). New Delhi, India, Apr.

(29) _____
1969. Supply of and Demand for Selected Agricultural Products in India, Revised Projections to 1980-81. New Delhi, India. Apr.

(30) International Wheat Council
Review of the World Grains Situation. London. Selected issues.

(31) _____
World Wheat Statistics. London. Published annually.

(32) Johnson, A.A. and Eapen, K.E.
1970. A Review of India's Package Program. Foreign Agriculture, U.S. Dept. Agr., Sept.

(33) Johnson, Sherman E.
1968. The IADP Path to Intensive Agricultural Development. Unpublished Report. Jan.

(34) Kriesberg, Martin
1969. Miracle Seeds and Market Economies. Columbia Journal of World Business, vol. IV, no. 2. Apr.

(35) Kronstad, Warren E.
1969. Global Crop Paper, Wheat. Spring Review of the New Cereal Varieties. Agency for International Development. May.

(36) Ladejinsky, Wolf
1970. Ironies of India's Green Revolution. Foreign Affairs, July.

(37) Leftwich, R.H.
1961. The Price System and Resource Allocation. Holt, Rinehard and Winston, New York.

(38) Mellor, John W.
1969. The Role of Government and the New Agricultural Technologies. Cornell University, Ithaca, New York, May.

(39) Mellor, John W. and Dar, Ashok K.
1968 Determinants and Development Implications of Foodgrains Prices in India, 1959-64. Am. J. Agr. Econ. 50:4. Nov.

(40) Paarlberg, Don
    1970. The World That Won Nobel Peace Prize for Iowan. Des Moines Sunday Register, Des Moines, Iowa, Oct.

(41) _____
    Peace is a Grain of Wheat: The Story of Nobel Peace Prize Winner, Norman E. Borlaug, War on Hunger. vol. V, no. 12, U.S. Dept. State (USAID), Dec.

(42) Pearson, L.B.
    1970. Can the Food Battle be Won? World Agriculture, International Federation of Agricultural Producers, Washington, D.C., Oct.

(43) Quisenberry, K.S., and Reitz, L.P. (editors)
    1967. Wheat and Wheat Improvement. American Soc. of Agronomy, Inc., Madison, Wisconsin.

(44) Reitz, Louis P.
    1968. Short Wheats Stand Tall. 1968 Yearbook of Agriculture, U.S. Dept. Agr.

(45) _____ and Salmon, S.C.
    1968. Origin, History and Use of Norin 10 Wheat. Crop Science. vol. 8, Nov. - Dec.

(46) Rice, Edward B.
    1970. Spring Review of the New Cereal Varieties - A Prospect, Agency for International Development, U.S. Dept. State, Mar.

(47) Rockefeller Foundation
    1959. Mexican Agricultural Program, 1958-59. New York.

(48) _____
    1964. Progress in the Agricultural Sciences, Annual Project 1963-64. New York.

(49) _____
    1968. President's Five-Year Review and Annual Report 1968. New York, Apr.

(50) Smith, John Newton
    1968. Argentine Agriculture, U.S. Dept. Agr., Econ. Res. Serv., ERS-Foreigh-216. July.

(51) Streeter, Carroll P.
    1967. Farming in the Future. Newsweek Globe Report, Aug.

(52) _____
    1969. A Partnership to Improve Food Production in India. A Special Report from Rockefeller Foundation, New York, Dec.

(53) Tsu, Sheldon K.
1969. The Long and Short of Semidwarf Wheat Varieties in Developing Countries, Discussion Paper No. 20, Unpublished, U.S. Dept. Agr., Econ. Res. Serv., July.

(54) United Nations Food and Agriculture Organization
1967. Agricultural Commodities--Projections for 1975 and 1985. vols. I and II. CCP, Aug.

(55) _____
1969. Production Yearbook. Rome.

(56) _____
World Grain Trade Statistics, Exports by Source and Destination. Rome. Published annually.

(57) U.S. Agency for International Development
1967. Gross National Product: Growth Rates and Trend Data by Region and Country. RC-W-138, Mar.

(58) U.S. Agency for International Development Mission, India
1969. Rice and Wheat in India. Spring Review of the New Cereal Varieties, New Delhi.

(59) _____, Pakistan
1969. Rice and Wheat in Pakistan. Spring Review of the New Cereal Varieties. Rawalpindi.

(60) _____, Turkey
1969. Introduction of Mexican Wheat into Turkish Agriculture. Spring Review of the New Cereal Varieties. Ankara.

(61) _____
1969. Wheat in Turkey, Spring Review of the New Cereal Varieties. Ankara, Mar.

(62) _____
1969. Wheat and Corn in Mexico. Spring Review of the New Cereal Varieties, ID/FAS/Dept. of Agr.

(63) _____
1969. Spring Review of the New Cereal Varieties, Latin America, Washington, D.C., Apr.

(64) _____
1969. Spring Review of the New Cereal Varieties--Implications for AID, May.

(65) _____
1969. The Role of Research, Spring Review Cereal Varieties, Washington, D.C., May.

(66) _____
    1969. Spring Review of the New Cereal Varieties--The Role of Institutions, Washington, D.C., May.

(67) _____
    1969. Spring Review of the New Cereal Varieties--Major Physical Inputs, Washington, D.C., May.

(68) _____
    1969. Spring Review of the New Cereal Varieties--Management System, Washington, D.C., May.

(69) _____
    1969. Spring Review of the New Cereal Varieties--Priorities and Programming, Unpublished, Washington, D.C., May.

(70) _____
    1969. Spring Review of the New Grain Varieties, Wheat, Unpublished, Morocco, Mar.

(71) _____
    1969. General Project Report, Morocco 1968-69, Unpublished, USAID/Rabat, Oct.

(72) _____
    1970. General Project Report, Morocco 1969-70, Unpublished, USAID/Rabat, Oct.

(73) _____
    1969. Wheat in Morocco. Spring Review of the New Cereal Varieties. AID/Rabat.

(74) U.S. Department of Agriculture
    1968. Food Grain Statistics Through 1967, Wheat, Rye, Rice, Flour, Byproducts., Stat. Bul. 423. Apr.

(75) _____
    1969. Indices of Agricultural Production in 1959-68 in Africa and the Near East., ERS-Foreign 265. Published annually.

(76) _____
    1969. Indices of Agricultural Production for East Asia, South Asia and Oceania. Average 1957-59 and Annual 1959-68. Published annually.

(77) _____
    1969. The Wheat Situation. WS-210, Nov.

(78) _____
    1971. The Wheat Situation. WS-215, Feb.

(79) _____
   1964. World Food Budget 1970. FAER-19.

(80) _____
   1970. Whither Wheat in the LDC's? Farm Index. Econ. Res. Serv., Oct.

(81) _____
   1969. Agricultural Statistics 1969.

(82) _____
   1951. Crops in Peace and War. The Yearbook of Agriculture, 1951.

(83) _____
   1936. Improvement in Wheat: The Yearbook of Agriculture, 1936.

(84) _____
   1968. Science for Better Living. The Yearbook of Agriculture, 1968.

(85) _____
   1970. Competition for World Wheat Markets and U.S. Exports, FAS M-214, Feb.

(86) _____
   1969. Wheats of World Commerce. Foreign Agriculture, Jan. 13.

(87) _____
   1971. World Agricultural Production and Trade Statistical Report, Mar.

(88) _____ Office of the Agricultural Attache, American Embassy
   1970. Brief on Indian Agriculture, Unpublished, New Delhi, India.

(89) _____
   1971. Agricultural Situation, American Embassy, New Delhi, India, Jan.

(90) _____
   1970. Grain and Feed, American Embassy, Ankara, Turkey, Nov.

(91) _____
   1971. Grain and Feed, American Embassy, Rabat, Morocco, Feb.

(92) U.S. House of Representative Committee on Foreign Affairs
   1969. The Green Revolution, Ninety-first Congress, First Session, Dec.

(93) Wall Street Journal
   1970. Argentine Wheat Production in 1971, Dec.

(94) West, Quentin M.
   1969. The Revolution in Agriculture: New Hope for Many Nations, 1969 Yearbook of Agriculture, U.S. Dept. Agr.

(95) Willett, Joseph W.
   1969. The Impact of New Grain Varieties in Asia. U.S. Dept. Agr., ERS-Foreign 275. July.

APPENDIX

Appendix Table 1.--India: Area, production, and yield of wheat, annual 1966-1970 and averages 1961-65 and 1968-70

Year of harvest	Area	Index	Production	Index	Yield	Index
	1,000 ha.		1,000 tons		Kg./ha.	
1961-65	13,402	100	11,202	100	836	100
1966	12,656	94	10,424	93	824	98
1967	12,838	96	11,393	102	887	106
1968	14,998	112	16,540	148	1,103	132
1969	15,958	119	18,651	167	1,169	140
1970	16,626	124	20,093	179	1,209	150
1968-70	15,861	118	18,428	165	1,162	139

Source: Table 2, reference (18).

Appendix Table 2.--Pakistan: Area, production, and yield of wheat, annual 1966-70 and averages 1961-65 and 1968-70

Year of harvest	Area	Index	Production	Index	Yield	Index
	1,000 ha.		1,000 tons		Kg./ha.	
1961-65	5,065	100	4,203	100	830	100
1966	5,210	103	3,952	94	759	91
1967	5,417	107	4,394	105	811	98
1968	6,061	120	6,477	154	1,069	129
1969	6,045	119	6,711	160	1,110	134
1970	6,349	125	7,399	176	1,165	140
1968-70	6,152	121	6,862	163	1,115	134

Source: Table 2, reference (18).

Appendix Table 3.--Turkey: Area, production, and yield of wheat, annual 1966-1970 and averages 1961-65 and 1968-70

Year of harvest	Area	Index	Production	Index	Yield	Index
	1,000 ha.		1,000 tons		Kg./ha.	
1961-65	6,807	100	7,050	100	1,036	100
1966	7,163	105	8,200	116	1,145	105
1967	7,204	106	9,000	128	1,249	121
1968	7,304	107	8,400	119	1,150	111
1969	7,650	112	8,300	118	1,085	105
1970	7,560	111	8,000	113	1,058	102
1968-70	7,505	110	8,233	117	1,100	106

Source: Table 2, reference (18).

Appendix Table 4.--Area and production of local and HYV wheat varieties as a percentage *
of total wheat for selected countries, annual, 1966-70

	1966		1967		1968		1969		1970	
	Area	Production	Area	Production	Area	Production	Area	Production	Area	Production
					Percent					
India										
Local	100	100	96	89	80	53	70	40	63	32
HYV	n.a.	n.a.	4	11	20	47	30	60	37	68
Pakistan										
Local	100	100	98	95	84	65	61	41	55	41
HYV	n.a.	n.a.	2	5	16	35	39	59	45	59
Mexico										
Local	---	---	---	---	---	---	---	---	---	---
HYV	100	100	100	100	100	100	100	100	100	100
Turkey										
Local	100	100	100	100	98	93	92	80	92	78
HYV	n.a.	n.a.	n.a.	n.a.	2	7	8	20	8	22
Afghanistan										
Local	100	100	100	100	99	98	96	89	95	86
HYV	n.a.	n.a.	n.a.	n.a.	1	2	4	11	5	14
Tunisia										
Local	100	100	100	100	100	100	98	93	93	73
HYV	n.a.	n.a.	n.a.	n.a.	n.a.	n.a.	2	7	7	27
Iran										
Local	100	100	100	100	100	100	100	99	98	97
HYV	n.a.	n.a.	n.a.	n.a.	n.a.	n.a.	n.a.	1	2	3
Morocco										
Local	100	100	100	100	100	100	100	100	98	96
HYV	n.a.	n.a.	n.a.	n.a.	n.a.	n.a.	n.a.	n.a.	2	4
Total										
Local	98	95	96	90	88	71	79	57	75	52
HYV	2	5	4	10	12	29	21	43	25	48

* Some percentages indicate 100 resulting from rounding. Note: n.a. indicates not applicable or not available.
Source: Table 2.

ACCELERATING INDIA'S FOOD GRAIN PRODUCTION

1967-68 TO 1970-71

REQUIREMENTS AND PROSPECTS
FOR A YEARLY GROWTH RATE
OF 5 PERCENT

FOREIGN AGRICULTURAL
ECONOMIC REPORT NO. 40

U.S. DEPARTMENT OF AGRICULTURE
ECONOMIC RESEARCH SERVICE

PREFACE

This report was first written for use by the U.S. Agency for International Development (AID) Mission to India in its program evaluation and planning during the summer of 1967. It was prepared in conjunction with papers on other subject matter areas, which together provided a comprehensive and fairly well-balanced analysis of India's agricultural production, potentials, and prospects. This report is presented essentially as first written, however, in the belief that both its substantive features and its methods of approach may be of interest to others concerned with the food problems of the world.

On the substantive side, this report indicates that India is on the move with respect to long needed improvements in agriculture, after having passed through the worst two consecutive drought years of this century. On the methodological side, it presents an approach to shortrun agricultural production projections of the kind frequently needed in international program operations which merits consideration for both its usefulness and its simplicity.

The authors are indebted to many people in the AID Mission to India as well as to persons in the Rockefeller Foundation, the Ford Foundation, the Ministry of Food and Agriculture of the Government of India, and other agencies for assistance in the preparation of this paper. The authors alone, however, bear full responsibility for choice of the data and information used in this report and for the interpretations that are made of them.

The Agency for International Development financed the research on which this report is based.

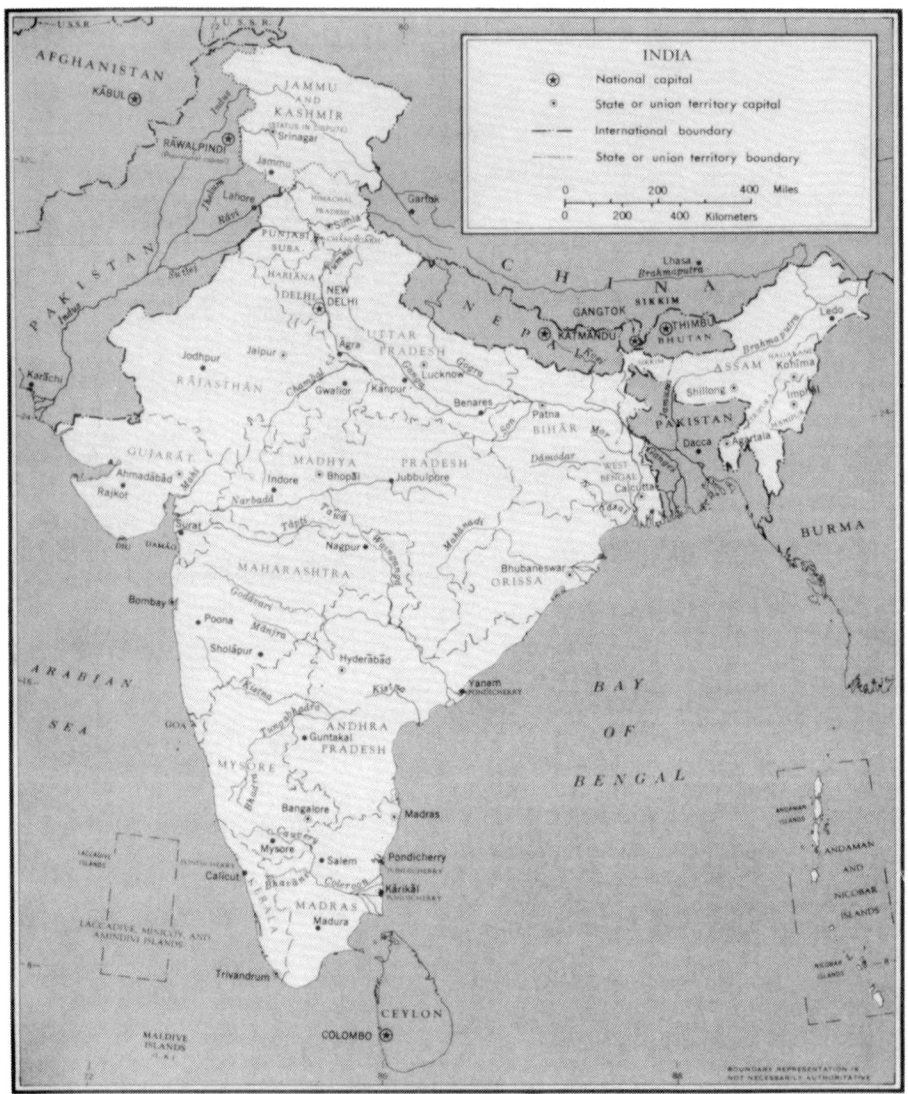

CONTENTS

	Page
Summary	iv
Introduction	1
Food Grain Production Trends	2
Output	2
Inputs	4
Directions of Policies and Programs	5
Recent Improvement in Food Grain Production Potentials	6
Technological Advances	6
Rice	7
Wheat	9
Bajra, Maize, and Jowar	9
Shifts in Policy	9
Estimation Model for 1967-68 Food Grain Output	10
Methodology	10
Inputs	11
Results	11
Requirements for a 5-Percent Growth Rate, 1967-68 to 1970-71	13
Inputs	13
Results	13
Policies and Programs	15
High-Yielding Varieties	16
Fertilizers	16
Irrigation	17
Plant Protection Materials	17
Transport Facilities	18
Agricultural Credit	18
Agricultural Research and Education	19
Incentives	19
References	20
Appendix	21
Comments on a Report of the President's Science Advisory Committee, The World Food Problem	21
Model for Estimating Food Grain Output	22
Appendix Tables	23

Washington, D.C.   March 1968

SUMMARY

This report deals with the potentials and requirements for increasing India's food grain production by 5 percent yearly from 1967-68 to 1970-71. It presents a model for projecting output by measuring the marginal product resulting from increases in farm inputs with response ratios based on tests and actual field cuts.

A 5-percent annual growth rate was chosen because it is near the minimal level needed to achieve self-sufficiency in food grain production within the next decade and it appears to be attainable and economically feasible. The year 1967-68 was chosen as a base since it holds promise as the turning point in India's agriculture.

The outlook for India's agriculture has greatly improved as a result of the introduction of new high-yielding grain varieties and sharply increased supplies of fertilizer and other farm inputs. These breakthroughs have occurred in only the past 3 or 4 years and come at a time when agriculture has been rocked by two consecutive years of drought--the most severe of the century. It is not coincidental that these advances were made in this period, for the importance of agriculture to India's economic progress has never been so dramatically illustrated as it was with the two poor grain crops of 1965-66 and 1966-67. This has led to increased emphasis upon policies and programs to accelerate expansion of agricultural output in India.

The key elements in India's improved agricultural base have been varietal breakthroughs for rice, wheat, maize, jowar, and bajra. These new high-yielding varieties are not only superior to native varieties under normal monsoon conditions but they greatly excel in their capacity for productively using fertilizer, water, and other inputs. These new grain varieties have been introduced in India in only the past 3 or 4 years and commercial adoption has expanded rapidly.

Fertilizer consumption in India has tripled in only 2 years as a result of increased imports and domestic production. This reflects changes in Government policies and programs for budget allocation, foreign investments, and particularly foreign exchange allocation.

Using estimates for the 1967-68 availabilities of high-yielding grain varieties, fertilizer, and irrigation, a model is developed which projects 1967-68 food grain production at 93.6 million tons which falls very near the level of the long-term annual trend of 2.7 percent. To reach an annual growth rate of 5 percent from 1967-68 to 1970-71 will require a substantial acceleration of the input base--fertilizer, pesticides, improved seed, and the like.

The model is used to find what base would be required to reach this growth objective in 1970-71. One base would include:

...121 million hectares sown to food grains

...38 million hectares irrigated for food grains

...13.2 million hectares sown with high-yielding varieties

...2.7 million tons of plant nutrients

These levels of inputs could be attained and, in fact, could be exceeded. So, the 5-percent growth objective is well within reach.

In the framework of the model is the assumption that the growth of India's agro-industry will be adequate to service the rising demands of agriculture. For example, farmers will need assured market outlets at incentive prices; marketing and storage facilities will need to be improved. It is recognized however, that there will inevitably be many day-to-day problems in this sector which must be solved for agriculture to successfully attain the desired rate of growth.

ACCELERATING INDIA'S FOOD GRAIN PRODUCTION 1967-68 to 1970-71

Requirements and Prospects for a Yearly Growth Rate of 5 Percent

by

William E. Hendrix, James J. Naive, and Warren E. Adams[1]

INTRODUCTION

This report deals with the potentials and requirements for increasing India's food grain production by 5 percent per year from 1967-68 to 1970.[2] It is composed of five major sections as follows:

...review of India's agricultural record since 1949-50,
...description of recent changes in technologies and policies providing a basis for accelerating growth,
...estimation model of 1967-68 food grain output,
...estimation of inputs and other requirements (within specified constraints) for a 5-percent growth rate, and
...review of current policies and programs bearing on the above requirements.

The year 1970-71 is the end of the fourth 5-year plan period. As such, it is the year toward which India's official targets on production, inputs, and other requirements are pointed.

The year 1967-68, instead of earlier years in the fourth plan period, is chosen as base for a yearly 5 percent takeoff in this report because
...1965-66 and 1966-67 were among the most severe drought years experienced by India in a century,
...1967-68 holds promise as a major turning point in India's food grain production potentials and in effectiveness of policies and programs for their realization.

Improvements made in India's agricultural base, particularly irrigation, since gaining independence in 1947 helped to cushion the adverse effects of the 1965-66 and 1966-67 droughts. Nonetheless, output dropped from 1964-65 to 1965-66 by the largest percentage for any year since 1920-21. The combined shortfall for 1965-66 and 1966-67 was larger than that for any other two consecutive years in this century.

These large production declines have provided dramatic illustration and created increased appreciation of agriculture's importance to India's general economic progress. This is reflected in greatly increased emphasis upon India's agriculture in the policies and programs of both Central and State governments, as well as of AID and other national and international development agencies.[3]

---

[1] William E. Hendrix, Agricultural Economist, Foreign Development and Trade Division, Economic Research Service, is in India engaged in research on factors associated with differences and changes in agricultural output and productivity. James J. Naive, Agricultural Economist, is with the Foreign Regional Analysis Division, Economic Research Service. Warren E. Adams was Economic Advisor to the Agricultural Division, AID Mission to India; he is now Professor of Economics, Earlham College, Richmond, Ind.

[2] The term "food grains" as used collectively in this report includes rice in milled equivalent and pulses. In contrast to most countries, in India grain is not generally used for livestock feed. In this report, "grain" will refer only to food grain. India's official food grain statistics are compiled on a crop-year basis which includes crops harvested during the 12-month period from July 1 of 1 year to June 30 of the following year. Thus, 1967-68 food grain production refers to those crops harvested in the last half of 1967 and the first half of 1968.

[3] India's government at the national level is commonly called the "Central Government" or simply "The Center" as is used later in this report.

Fortunately for the likely success of this new emphasis, it closely parallels large recent advances in adaptable farm technology in India which some believe have more to offer than all the other farm technological advances put together in the first half of this century. The key elements of these advances consist of varietal breakthroughs for India's major cereal crops-- rice, wheat, jowar, bajra, and maize.[4] These hold promise of yield increases roughly comparable to that recently achieved for hybrid corn in the United States. The importance of such gains for India seems particularly great because of the large relative importance of cereals in total agricultural production.[5]

A food grain production growth rate of 5 percent per year has been chosen for the purposes of this analysis because:
...it is near the minimal level needed by India to achieve its own stated objective of self-sufficiency in grain production within the next decade,
...it appears to be attainable and economically feasible, assuming appropriate policies and programs for providing inputs, supporting facilities and services, and incentives.

From the side of needs, India must increase its grain production by 2.5 percent per year (some estimates run to 2.7 percent) just to feed its growing population at present per capita consumption levels and at the current level of self-sufficiency.

An additional increase of 1 percent or more per year is needed to meet increases in demand expected from rising per capita incomes.

Finally, an additional rate of increase in output is needed
...for progress toward India's stated goal of self-sufficiency in grain production;
...for replenishing now exhausted contingency stocks of grains, normally held by farmers, traders, and nonfarm households; and
...for building buffer stocks to stabilize market supplies and prices.

Fortunately, the rate of growth required to meet the last three needs turns upon India's own sense of urgency. For at least the next 3 to 5 years, India can effectively absorb as large an increase in food grain production as it can economically produce.

A 5-percent annual rate of growth from 1967-68 to 1970-71 would be a sharp upturn from historical rates of growth. It will meet the needs from population and per capita income growth and enable India to move toward its goal of grain and general economic self-sufficiency.

FOOD GRAIN PRODUCTION TRENDS

Output

India's progress in increasing food grains since independence has fallen short of its goals and needs. It is instructive, however, to look at its record:
...in context of the political, social, and economic problems that India as a new nation has faced; and
...against progress prior to Independence.

India's main problem since Independence has been that of integrating under a new, democratic nation a population--
...larger than that of the whole western hemisphere; larger also than that of all of Europe outside the USSR;
...more impoverished and illiterate than that of any but a few relatively small Asian and African countries;
...more diverse in ethnic features, languages, religions, and political ideologies than is that of the whole population of Europe.

Even so, India's grain production record since independence looks good compared with that of the preceding half century. The production record in the first half of the 20th century for the area now comprising both India and Pakistan is as follows (21):

Time Period	Annual Average
	(Million tons)
1900-01 to 1909-10	67.6
1910-11 to 1919-20	72.7
1920-21 to 1929-30	68.1
1930-31 to 1939-40	67.8
1940-41 to 1947-48	67.4

In contrast, from 1949-50 to 1964-65, India as now constituted increased its food grain production by an average of nearly 2 million metric tons per year (table 1).[6]

---
[4] Jowar is the Indian term for grain sorghum; bajra is spiked or pearl millet; and maize is corn.
[5] Food grains as used in this report account for about 75 percent of India's gross agricultural production. For an excellent report on Indian agriculture, see (13).
Underscored numbers in parentheses refer to references at the end of this report.

[6] Unless noted otherwise, tons are metric.

Table 1.--India: Food grain production, 1949-50 to 1966-67 and "Trend" estimates of production, 1967-68

Year	Actual output	Moving averages of output, 1949-1950 to 1964-65--	
		3-year	5-year
	----------------Thousand metric tons----------------		
1949-50.........	60,653	--	--
1950-51.........	54,922	57,028	--
1951-52.........	55,508	57,368	60,988
1952-53.........	61,673	63,122	62,979
1953-54.........	72,186	68,155	65,838
1954-55.........	70,606	70,669	69,204
1955-56.........	69,216	70,720	70,170
1956-57.........	72,337	69,352	71,470
1957-58.........	66,504	72,509	72,689
1958-59.........	78,687	73,963	75,249
1959-60.........	76,699	79,135	77,323
1960-61.........	82,018	80,474	79,712
1961-62.........	82,706	81,057	80,023
1962-63.........	78,448	80,466	82,482
1963-64.........	80,243	82,562	--
1964-65.........	88,996	--	--
1965-66.........	72,030	--	--
1966-67.........	75,049	--	--
1967-68 trend estimate[1].....	--	[2] 95,730	[3] 93,940

[1] Omits use of 1965-66 and 1966-67 data.
[2] Using 1957-58 as "Origin" for computational purposes, $Y = 72416 (1.0283)^t$ where Y = output, and t = time in years.
[3] With 1957-58 as "Origin", $Y = 72038 (1.0269)^t$.

Source: (9).

Calculated on the basis of its annual output series, unadjusted for weather and associated yield variations, India had an output growth rate of 2.98 percent (compound) per year. Using moving averages to smooth out irregularities caused by weather, it had a growth rate of 2.83 percent using a 3-year average and 2.69 using a 5-year average. Projecting 1967-68 output at trend growth rates of 2.83 and 2.69 percent indicates an output of 95.7 million and 93.9 million tons, respectively.

Neither the 3-year nor the 5-year moving averages show a marked slowdown in the grain growth rate between the first and second half of the 1949-50 to 1964-65 period, such as is indicated from use of the unadjusted output data. The 5-year moving average indicates consistent year-to-year increases and a nearly imperceptible decline in rate of growth. Even for such decline as is indicated, one cannot be wholly sure whether it reflects a genuine shift in trend or is only the result of using a period of time too short for even a 5-year moving average to smooth out the influence of weather fluctuations that are quite normal to India.

Large shortfalls in production in 1965-66 and 1966-67 resulting from highly abnormal weather have focused world attention on India's food problem and created the impression that India's agriculture is nearly stagnant while its population is increasing by 2.5 percent or more per year.

India's agriculture has always been subject to large year-to-year variations in output as a result of the variable and uncertain monsoon rains upon which it depends. It has experienced severe famine extending over large parts of one or more of its major regions many times in its history. Twenty-seven famines, each extending over areas equal in size to one or more States such as Gujarat and Orissa, occurred in the 19th century. Many of India's droughts before 1900, however, resulted in famine, largely because of poor transport and communication facilities and lack of administrative machinery for procurement and distribution from surplus to deficit areas.

Since 1900, famines have occurred less frequently. India has, however, experienced an annual drop of 10 percent or more in its grain production five times since 1900. These years and the associated percentage declines in output were as follows (21):

Year	Percent
1907-08	12.9
1918-19	32.3
1920-21	24.0
1923-24	16.6
1965-66	18.8

Since 1923-24, famine or near-famine conditions resulting from drought have occurred much less frequently than between 1900 and 1923-24. However, frequent declines in output of less than 10 percent per year have continued to characterize Indian agriculture. In the 15-year period between 1949-50 and 1964-65, the following six declines occurred (in thousand metric tons):

From 1949-50 to 1950-51	5,731
From 1953-54 to 1954-55	1,580
From 1954-55 to 1955-56	390
From 1956-57 to 1957-58	5,833
From 1958-59 to 1959-60	1,988
From 1961-62 to 1962-63	4,262
Total	19,784

From 1964-65 to 1965-66, India's grain production dropped by 16,732,000 tons as a result of widespread drought. This was a shortfall equal to 85 percent of the sum of the above six annual declines occurring between 1949-50 and 1964-65. Worse still, this was followed by a second severe drought in 1966-67 in Bihar, eastern Uttar Pradesh, large parts of Madhya Pradesh, and parts of other States, most of which have dense populations and normally productive land.

That the recurrence of severe drought and near-famine conditions in 1965-66 and again in 1966-67 is the prelude to a new weather cycle and production declines of the frequency and magnitude experienced between 1800 and 1923-24 is highly doubtful--if for no other reason than that India now has chose to 40 million hectares of land under irrigation.

But whatever the frequency of droughts like that of 1965-66, even mere year-to-year output fluctuations of the frequency and extent of those between 1949-50 and 1964-65 make it difficult to obtain a statistically reliable estimate of India's rate of growth in food grain production from observations covering only 5 to 6 years such as from 1958-59 to 1963-64. Even for periods of 15 to 20 years, one needs to take careful account of yearly fluctuations caused by weather. This is attempted in this report by the use of 3-year and 5-year moving averages.

However for 1965-66 output, even a 5-year moving average differs substantially from the trend of earlier years or a 1965-66 projection based upon available inputs and normal output response ratios.

Data on output by States indicate that a few States had a larger output in 1966-67 than in 1964-65, notwithstanding somewhat less favorable weather in 1966-67 (tables 8 and 9).

Inputs

Inputs of land, irrigation water, labor, and fertilizers used in India's agriculture have been increasing rather steadily since 1950-51 (table 2). Gross sown area, however, only increased from 156.1 million hectares in 1961-62 to 157.9 million hectares in 1964-65. However, from 1960-61 to 1961-62, it increased by 3.4 million hectares after two earlier years of very little change.

Compensation for this slowdown in area growth, however, has been provided in large part by increases in irrigation, fertilizers, and other inputs. From 1952-53 to 1964-65, total fertilizer consumption in terms of plant nutrients increased tenfold, or by 586,880 tons. This is an amount sufficient to yield an increase in food grain output of 3.8 million tons, assuming a response ratio of 6.5. This output would equal that from about 5 million hectares of land at average yield levels. Fertilizer consumption in 1967-68 is expected to reach 2.1 million tons, enough over the 1964-65 level to yield an output equal to what might be expected from the addition of 16 million hectares of land. Nitrogen

Table 2.--India: Major agricultural inputs, 1950-51 to 1967-68[1]

Year	Major inputs			
	Land[2]	Water[3]	Labor[4]	Fertilizer[5]
	Thousand hectares	Thousand hectares	Thousand agr. workers	Metric tons
1950-51	131,893	22,563	102,929	---
1951-52	133,234	23,180	103,217	---
1952-53	137,675	23,305	103,506	65,685
1953-54	142,480	24,363	103,796	104,803
1954-55	144,083	24,948	104,087	120,934
1955-56	147,311	24,642	104,429	130,777
1956-57	149,492	25,707	104,789	153,719
1957-58	145,832	26,628	105,149	183,727
1958-59	151,629	26,948	105,509	223,844
1959-60	152,824	27,413	105,869	304,598
1960-61	152,716	27,886	106,186	293,871
1961-62	156,099	28,373	106,505	383,450
1962-63	156,736	29,452	106,824	477,921
1963-64	156,970	30,380	107,144	574,220
1964-65	157,940	31,170	107,465	652,565
1965-66 (estimate)	---	---	---	757,287
1966-67 (estimate)	---	---	---	1,320,000
1967-68 (estimate)	---	---	---	2,250,000

[1] Includes inputs used on other crops as well as on food grains.
[2] Gross sown area.
[3] Gross irrigated area.
[4] Agricultural workers as reported in National Income Account reports for selected years and estimated for intervening years using rates of change indicated in National Income Accounts Statistics.
[5] Tons of plant nutrients (N, P_2O_5, and K_2O).

Source: (2), (8), and (10).

consumption alone in 1967-68 will reach the total attained in the United States in the early 1950's.[7]

Multiple-cropping is an additional way of extending the effective land area. At present, only one crop per year is raised on 85 percent of India's net sown area. Much of the double-cropping is done on unirrigated land. Only about 15 percent of the net irrigated area is being used for 2 or more crops per year. With assured supplies of water the year round, two to three crops per year can easily be grown under Indian climatic conditions.

[7] The total cropped area in India, which takes into account multiple-cropping (land producing more than 1 crop per year), is approximately equal to that in the United States. Thus comparison of total nitrogen consumption for the two countries is valid.

Directions of Policies and Programs

In early efforts to modernize India's agriculture following independence, it was widely assumed that the technology for doing so was readily available; these efforts consisted of applying:

...indigenous techniques already employed by the better farmers, and

...importable technologies originally developed for farmers of economically advanced nations.

Emphasis in these earlier efforts, therefore, centered heavily upon building new institutions to facilitate adoption of known technologies rather than upon strengthening technological bases. These included:

...extension activities built around widespread use of village-level workers and community development programs,

...cooperatives to provide credit, and to distribute fertilizers, seeds, and other supplies,

...land reform to provide incentives to India's millions of tenants to adopt better methods, which under existing tenurial arrangements, would increase output but not their income.

Such price policies as were in effect before the 1960's were directed more to consumer interests than to larger incentives and smaller price risks for producers. Terms of trade (prices) between food grains and nonagricultural commodities therefore shifted through most of the 1950's in favor of the latter, to the detriment of farmers and agriculture as an industry.

The foregoing policies among States and smaller areas of India have met with varying degrees of success within the limits of available technologies. Agricultural output in Punjab (as constituted in 1965), Gujarat, and Madras increased from 1952-53 to 1964-65 by a compound rate of more than 4 percent per year (table 11). In four districts in the Punjab and two in Madras State, agricultural production increased on average more than 7 percent per year.

These high rates of growth reflected the presence of determined agricultural leadership which was above average in initiative, decision-making, and administrative experience. This leadership has been successful in assisting farmers in these areas to obtain more fertilizers, more irrigation facilities, and more technical assistance. Such leadership often is found in areas where the spirit of enterprise and entrepreneurial abilities are most widely developed. Some observers have noted that in India's more rapidly developing States, agriculture has been organized in large part around owner-operator freeholds, in contrast to large land-holding estates such as are found in the slow-growth State of Uttar Pradesh.

RECENT IMPROVEMENT IN FOOD GRAIN PRODUCTION POTENTIALS

The achievement of a 5-percent annual growth rate in national food grain production requires increasing the rate throughout most of India to the levels that a few States and, in particular, a few districts within these States have demonstrated is technically possible. The basis for doing this has been greatly improved as a result of recent developments in the following two important aspects of the Nation's agricultural economy: (1) Applicable farm technology and (2) policies and programs of both Central and State governments directed to the adoption of technological improvements.

Technological Advances

The key element in India's recent farm technological advance consists of highly productive varietal breakthroughs for rice, wheat, maize, jowar, and bajra. Supplies of new high-yielding varieties are large enough to insure relatively large increases in 1967-68 plantings.

A somewhat comparable technical advance in U.S. agriculture was the development and commercial adoption of high-yielding hybrid corn. After these were first successfully adopted in the Corn Belt in the 1930's, however, it took more than another decade of further research in other regions to develop hybrids well adapted to their soil and climatic conditions. In the United States, similar varietal advances for wheat, grain sorghums, and other cereals came several years later.

In contrast to the U.S. case, new highly productive varieties of rice, wheat, maize, jowar, and bajra have all come into commercial use in India within only the last 3 to 4 years, as a result of the transferability of varieties produced elsewhere and of India's own research.

Before turning to available information on yields and other attributes of these new varieties, brief reference to India's traditional crop varieties will help to set these varietal breakthroughs in their proper perspective.

India's traditional crop varieties have evolved over centuries as the surviving species in a harsh physical environment. This environment has been marked by frequent extremes of droughts and floods, uncertain and widely varying moisture conditions, low soil fertility, and crude tillage practices plus other complex crop production and soil management problems characterizing tropical and semitropical regions.

The crop varieties that have evolved out of this harsh environment have been well adapted to it, especially in terms of survival capacities. Except under such extreme drought as that which recently occurred in Bihar, they have usually yielded a crop of some size when imported varieties have failed. They have, in other words, demonstrated a capacity for withstanding large variations in soil moisture and associated intake of plant nutrients without correspondingly large variations in yields. These have been exceedingly important qualities, contributing for centuries to the survival of Indian farm people.

On the other hand, the very genetic features that have enable these varieties to serve the needs of Indian agriculture so well in the past, lower their response to fertilizers, water, and other inputs. Indigenous varieties have shown relatively low response and capacity to absorb such inputs within economically profitable limits.

Moreover, until recently, even the improved varieties developed in temperate climatic zones have shown little adaptability to tropical and semitropical conditions or to latitudes other than those for which they were developed. One reason for this is their high sensitivity to variations in length of day and sunlight intensity. Hence, in countries like India, available crop varieties have functioned as severe constraints to increasing agricultural output except at costs much higher than those required for comparable output increases in the United States.

In the case of wheat, new high-yielding varieties whose genetic features make them insensitive to variations in sunlight and therefore easily adaptable within wide latitudinal ranges have recently been developed. Paralleling this work, there has been much effort under leadership of India's scientists, working closely with those of other nations, to develop hybrids well adapted to India.

These new varieties are not only superior to traditional varieties under normal monsoon conditions but they greatly excel local varieties in their capacity for using fertilizer, water, and other inputs. In fact, larger inputs of fertilizers and plant protection materials together with assured supplies of water cannot be overemphasized as essential to the continuing success of the high-yielding varieties. Expressed in another way, the new high-yielding varieties involve more than the mere substitution of one kind of seed for another. <u>Their successful introduction will require changes in nearly all components of Indian food grain production technology.</u>

Rice.-- Turning to specific varietal introductions, one rice variety now in fairly large-scale commercial production is ADT-27, which was developed in Madras State. In 1965, an average paddy yield of 3,820 pounds per acre was obtained on about 3,000 acres of ADT-27 grown under farm conditions in Tanjore District in the State of Madras. Yields ranged from 1,600 to 5,500 pounds with the top decile of growers having an average yield of 5,140 pounds and the lowest decile an average of 2,480. In 1966, under less favorable weather conditions and with the crop area increased to about 125,000 acres, the average yield of ADT-27 was 2,450 pounds. This was very favorable, compared with 1,760 pounds for "other improved varieties." Fertilizer use in the 1966 field trials was as follows:

Rice variety	Percentage of fields fertilized	Pounds of plant food per acre	
		Fields fertilized	All fields
ADT-27	97	68	64
Other Improved Varieties	80	47	37
Common Indigenous	75	37	28
Mixtures	55	29	16

Fertilizer yield responses for ADT-27 were somewhat low in 1966, probably because of unfavorable weather. But even then at up to 50 pounds of fertilizer per acre there was a response ratio of slightly over 28 to 1. The results were as follows.

Plant food (Pounds/acre)		Percentage of fields	Paddy yield
Group	Average	Percent	Pounds/Acre
0	0	3	1320
Under 50	33	38	2250
50-70	60	14	2400
70-90	80	23	2550
90-100	100	11	2810
110 & over	140	11	3080
Average	64	100	2450

Results of rice variety trials conducted in the 1966 kharif[8] season under auspices of the Indian Council of Agricultural Research with the Rockefeller Foundation cooperating are shown in table 3 for two levels of nitrogen application. In these trials, conducted in all areas of India, local Indica varieties not only had appreciably lower yields than did new Dwarf Indica and Ponlai varieties, but also demonstrated an appreciably lower response to fertilizers. In applications of nitrogen up to 50 kilograms per hectare, the response of improved varieties exceeded that of local varieties by more than 10 units of grain per unit of fertilizer used. This suggests a total response ratio of more than 20 to 1 for the improved varieties, for this range of nitrogen application.

[8] Fall and winter harvest season.

Table 3.--India: Summary of yields of specified rice varieties in the uniform variety trials, kharif 1966

Variety and type	Locations reporting	Yields of grain with nitrogen applied at--		
		50 kg/ha	100 kg/ha	Difference
	Number	Kg/ha	Kg/ha	Kg/ha
Dwarf Indica:				
TN-1 X Taichung 67	14	3,885	4,351	466
Taichung Native 1	20	3,603	4,319	716
Dee-Geo-Woo-Gen	15	3,644	3,899	255
IR 9-60	17	3,445	3,857	412
Ponlai:				
Kaohsiung 68	19	3,729	4,198	469
Tainan 3	20	3,577	4,155	578
Chianung 242	20	3,344	3,947	603
Taichung 65	18	3,543	3,884	341
Ch. 242 X CI 9155	17	3,128	3,479	351
Local Indica:				
NC 1626	14	2,893	3,200	307
Co 29	14	2,884	3,167	283

Source: (14).

Wheat[9].-- Preliminary releases by personnel working on the Intensive Agricultural District Program, the Farm Management Group, Ford Foundation, reveal the following results on wheat yields for the 1966-67 crop in Ludhiana District in Punjab State:

Variety and year	Yield (Lb./A.)
Mexican 1966-67	4,200
Indian 1966-67	2,130
All Varieties 1965-66	1,970
All Varieties 1964-65	2,015

It is estimated that Ludhiana had 37,000 acres of the Mexican dwarf wheat varieties in 1966-67, constituting 11 percent of its total wheat area. This was probably grown by better farmers, which partially accounts for a yield nearly twice as large as that obtained for Indian varieties. Yields of Indian varieties during the 3 seasons since 1963-64 have varied little. All of the farmers growing Mexican wheat used nitrogen fertilizers and 73 percent used phosphate in addition; the average applications were 84.5 pounds of N and 23.3 pounds of P_2O_5 per acre. The average application for all wheat (including Mexican) in the district was 53.6 pounds of N and 11.9 pounds of P_2O_5 per acre. Fertilizer use for the Mexican varieties exceeded that for the Indian varieties by about 48 pounds per acre; average yield of the Mexican wheat was 2,070 pounds higher. Thus the Mexican varieties yielded about 44 pounds of grain per additional pound of fertilizer. This high coefficient is the response to a whole complex of practices rather than to fertilizer alone. However, a response of 15 to 20 pounds of wheat per pound of fertilizer would seem reasonable for high-yielding varieties under average farm conditions.

Bajra, Maize, and Jowar.-- Data are available on varietal tests for bajra for 1965-66. In all test areas, yields for hybrids were higher than for local varieties. Even without fertilizer application, the average yields in one set of tests were 1,856 kilograms per hectare for local varieties compared with 2,154 for hybrids (table 4). The large advantage of the hybrids over local varieties, however, lies in their capacity to use larger amounts of fertilizers and to use them more productively. For example, the first increment of 40 kilograms of N resulted in yield increments of 713 kilograms for local varieties but 1,407 for hybrids, twice as much as for local varieties. Again, these results suggest response ratios of better than 15 to 1 for fertilizers used.

Tests conducted for 4 years on double-cross-hybrids of maize indicate grain yields of 3,300 to 7,000 kilograms per hectare (up to 100 bushels per acre). In all tests, yields of hybrids were much above those of local varieties, running generally 40 to 50 percent higher.

Available data on jowar indicates that yields for hybrids average about 500 kilograms per hectare higher than those for local varieties. Response ratios for varying ranges of nitrogen application were substantially higher for hybrid varieties as shown in table 5.

In evaluating the above test results, it should be emphasized that they have been obtained on better-than-average farms with better-than-average provision of technical assistance. They do, however, indicate potentials which may be reached as India's farmers gain experience and knowledge of the new high-yielding varieties and of their input and tillage requirements.

Shifts in Policy

Food crises in the last 2 years have had a dramatic impact upon the thinking of policymakers at all levels--Center, State and local--in matters pertaining to agriculture. Hence the commercial adoption of new high-yielding varieties and provision of assured water supplies, fertilizers, plant protection materials, and other inputs that are part of the new technology have been greatly facilitated by a new sense of urgency and determination to avert food crises like those of 1965-66 and 1966-67.

New directions of effort are being pointed directly to increasing production through more adequate provision of essential inputs in contrast to emphasis in the 1950's upon major institutional reforms. The wisdom of the current policy is reflected in the increased use of fertilizers, improved seeds, and other inputs and the fact that institutional impediments are not currently bottlenecks to the utilization of these inputs.

Current operative policies and programs are treated in fuller detail following the sections on 1967-68 output and requirements for a 5-percent growth rate, so as to better relate current and prospective achievements more directly to requirements.

---

[9] Data in this report are discussed in the terms that they are reported in statistics from India. Here wheat yields are discussed in terms of pounds per acre.

Table 4.--India: Yields of hybrid and local varieties of bajra at varying rates of nitrogen application, trial at Fatehabad (Agra) Uttar Pradesh, Kharif 1965

Nitrogen	Yields of grain		
	Local varieties	Hybrids	Differences
	----------------Kilograms per Hectare----------------		
0	1,856	2,154	298
40	2,569	3,561	992
80	3,069	4,348	1,279
120	3,806	5,645	1,839
160	3,393	5,967	2,574

Source: (15).

Table 5.--India: Response ratios of local and hybrid varieties of jowar for varying rates of nitrogen application

Variety	Response ratios for ranges of nitrogen application of--		
	0 to 40 Kg/ha.	0 to 80 Kg/ha.	0 to 120 Kg/ha.
	------Kilograms of Jowar per Kilogram of Nitrogen-------		
Local.........	14.2	4.8	--
Hybrid.......	19.2	16.1	13.0

Source: (17).

ESTIMATION MODEL FOR 1967-68 FOOD GRAIN OUTPUT

Although table 1 showed a trend extrapolation of output that would lead to a 1967-68 projection of about 95 million tons of food grain, forcasting production for a single year such as the current crop year depends upon the supply of inputs.

Methodology

An aggregative framework has been constructed for measuring the production response from these factors. Weather for this forecast is assumed to be normal.[10] In addition, it is assumed that relative prices are at levels which will provide cultivators the incentive to purchase the necessary inputs.[11]

The projection method used here measures the marginal product or output resulting from input changes from a base period to the period under review. The production responses from these input changes are based on likely input-output ratios, using fertilizer as the standard input.[12] This output added to the base period production results in the forecasted or projected output. This method has the

[10] Rainfall during the 1967-68 kharif and rabi seasons has been highly favorable.

[11] This also subsumes that credit is available when necessary for input purchases.
[12] See the discussion on "Recent Improvement in Food Grain Production Potentials."

advantage of taking into account any shift in the production function. The trend extrapolation, on the other hand, implicitly assumes no shift in the production function.

The base period used in this framework is the 3-year average centered on 1960-61. This period was selected for the following reasons: (1) Fluctuations in production caused by weather were relatively moderate; (2) fertilizer consumption was relatively low and the use of improved crop varieties was virtually nonexistent; (3) a projection base at the outset of the 1960 decade was convenient; and (4) it fitted the time references of previous projection studies (1) (12).

Inputs

The 1967-68 inputs for food grains used in this model are estimates based on targets of the Government of India; the self-help measures, as specified in Item V of the P.L. 480 agreement signed on February 12, 1967; and current reports on input supplies. They include the following:
...117.5 million hectares sowed to grains
...32.0 million hectares of gross irrigated grains area
...1.6 million tons of fertilizer in terms of plant nutrients nutrients applied to grains[13]
...6.1 million hectares sown with high-yielding varieties

Table 6 provides a comparison with the base-period inputs. In effect the model's task is to calculate the production response from incremental increases of 1.3 million sown hectares of food grains, 9.7 million gross hectares of irrigated area, 1.4 million tons of fertilizer, and 6.1 million hectares sown with high-yielding varieties.

Results

The model first accounts for the production increment attributed to only the increase in area, holding yields constant. This amounted to 885,000 tons, or 1.1 percent of the base-period production. Yields are held constant by increasing irrigation and fertilizer consumption at the same growth rate as area.

The next step estimates the increment resulting from the sowing of 6.1 million hectares of high-yielding grain varieties,

with the assumption that all of this area will be irrigated and fertilized at the rate of 60 kilograms per hectare. Thus 366,000 tons of fertilizers are applied to 6.1 million irrigated hectares of high-yielding grain varieties. A response coefficient of 13.5 was assumed, resulting in a production increment of 4.9 million tons.[14]

The third step measures the output increment from the unused irrigated area of 3.3 million hectares: Only local varieties would be sown; a fertilizer application rate of 40 kilograms per hectare is assumed, which would amount to 133,000 tons. A response coefficient of 9.0 is assumed which results in additional output of 1.2 million tons.

The residual input is 944,000 tons of fertilizer. This fertilizer is applied to nonirrigated land with local varieties of grains. A response coefficient of 6.5 is assumed which results in a production increment of 6.1 million tons.

The final step totals the production increments and the base-period production, resulting in an estimate of 93.6 million tons of grains in 1967-68. Thus, this analysis more than supports the trend projections of 94 to 95 million tons. The difference between the estimated 93.6 million and the 92 million set for the base should be regarded as a margin of safety for uncertainties of weather, input supplies and distribution, and response coefficients.

The assumption in the third step of applying residual fertilizer to nonirrigated land only is a conservative element of this model. It could be reasonably assumed that at least a portion of the fertilizer might be applied to the irrigated area utilized in step one (22.6 million hectares), after accounting for the area increase. As the model stands, only 136,000 tons of fertilizer or an average of 5.9 kilograms per hectare is applied to this area. If all of the remaining fertilizer (944,000 tons) were applied, then the rate would jump to 47.7 kilograms per hectare. If other things

---

[13] Including N, P_2O_5, and K_2O. Hereafter a unit of fertilizer will be assumed to contain 4 parts N, 2 parts P_2O_5, and 1 part K_2O. It is assumed that food grains account for 75 percent of total fertilizer consumption.

[14] Response as used in this context refers to the output resulting from a combination of inputs, but the coefficient will always refer to the fertilizer in the combination.

This is believed to be a fairly conservative response ratio. It is used because of an awareness of technical problems commonly encountered in the rapid spread of new crop varieties and other new practices. As India's farmers gain experience in use of new varieties, the response ratio can be expected to approach the levels now being obtained in experiments and in the Intensive Agricultural District Program (IADP) where reasonably good programs of technical assistance have been developed. (The IADP was initiated as a joint effort of the Ford Foundation and the Center. For a more detailed description see (11).)

Table 6.--India: Model for estimating 1967-68 food grain production

Inputs and outputs	Units	1959-60 to 1961-62 average[1]	Estimates for 1967-68[2]	Estimates of grain output increases from 1959-60 to 1961-62 (Ave.) to 1967-68 with following input increases[3]				
				Area	High yield varieties	Irrigation with local varieties	Fertilizer with non-irrigated local varieties	Total increases
		(1)	(2)	(3)	(4)	(5)	(6)	(7)
Inputs:								
Grain Area............	1,000 hectares	116,212	117,500	1,288	0	0	0	1,288
Gross irrigated Grain area..........	1,000 hectares	22,318	32,000	245	6,100	3,337	0	9,682
Fertilizer for Grain................	1,000 tons	131	1,575	1	366	133	944	1,444
High-yielding Varieties............	1,000 hectares	0	6,100	0	6,100	0	0	6,100
Output:								
Increments...........	1,000 tons	---	13,159	885	4,941	1,197	6,136	13,159
Total................	1,000 tons	80,465	93,624	---	---	---	---	---

[1] Irrigated grain area accounts for about 80 percent of total irrigated area. It is assumed that 40 percent of total fertilizer was applied to grain.
[2] Irrigated grain area accounts for 80 percent of total irrigated area. It is assumed that 75 percent of total fertilizer was applied to grains. Output expected with average wether conditions and with indicated inputs.
[3] Increases in grain output estimated as follows:
Column 3 - Yield held constant; production, irrigated area, and fertilizer increases at rate of area increase (1.1 percent).
Column 4 - High yield varieties grown on irrigated land and fertilized at 60 kg/ha; assumed yield response of 13.5 kg. grain for 1 kg. of fertilizer.
Column 5 - Fertilizer applied at 40 kg/ha; assumed yield response of 9 kg. grain for 1 kg. fertilizer.
Column 6 - Residual amount of fertilizer available assumed to have a yield response of 6.5 kg. grain for 1 kg. fertilizer.

are held constant, the output response from fertilizer is higher on irrigated land than on nonirrigated land (16). An increase in the response coefficient from 6.5 to 9.0 would then result in an additional 2.4 million tons of food grains.

If the input and production estimates for 1967-68 prove to be correct and output is merely near the trend level, it would suggest that the input base--fertilizers, high-yielding varieties, and irrigation--must be accelerated substantially over recent rates in order to reach a desired annual growth rate of 5 percent in the near future. The input base of 1967-68 is vastly improved from recent years, but apparently it will only substitute for the rapid expansion in area and increases in other production factors during the fifties in sustaining the historical growth rate.

REQUIREMENTS FOR A 5-PERCENT GROWTH RATE, 1967-68 TO 1970-71

At an annual growth rate of 5 percent from a 1967-68 estimated output of 92 million tons, India's grain production would reach 106 million tons in 1970-71. With this objective in view, the immediate task is to find what input base would be required to reach this output objective.

Inputs

For this computation the following assumptions were made:
...Normal weather will prevail;
...relative producer prices will be at levels which will provide cultivators the incentive to purchase and use the projected inputs;[15]
...the gross grain area will total 121 million hectares, 3 percent above the estimated 1967-68 level.[16] It is expected that part of this increase will be the result of multiple-cropping;
...the gross irrigated grain area will total 38.0 million hectares;[17]
...the area sown with high-yielding varieties will total 13.2 million hectares (the fourth plan target);

[15] This also subsumes that credit is available when necessary for input purchases.
[16] The area increase is taken as a trend extrapolation as projected in (20).
[17] Irrigated food grain area accounts for about 80 percent of total gross irrigated area.

...the area of high-yielding varieties will be irrigated and fertilized at the rate of 80 kilograms per hectare. The response coefficient is 13.5;
...fertilizers will be applied to the irrigated area with local varieties at the rate of 60 kilograms per hectare. The response coefficient is 9.0;
...an input-output coefficient of 6.5 for fertilizer applied to nonirrigated area with local varieties.[18]

The 1970-71 level of three input variables--land, high-yielding varieties and irrigation--has already been assumed or projected, simplifying the task of computing an input base. To compute the quantity of fertilizer necessary to reach 106 million tons, the model used to measure the marginal response of input increases is essentially the same as that used for the 1967-68 estimate. Again the base period is centered on 1960-61. The model must now find the necessary fertilizer, given other inputs and output, whereas for 1967-68 its assignment was to find output given the inputs.

Results

The computational steps follow the pattern of the 1967-68 input model as shown in table 7. The first calculation is the production increment resulting from the area increase (holding yield constant) of 4.8 million hectares; this amounts to 3.3 million tons. To hold yield constant requires 4,000 tons of fertilizer and 915,000 hectares of irrigated area in excess of the base-period levels.

The additional output resulting from the use of 13.2 million hectares of high-yielding varieties is computed in the second step; this totals 14.3 million tons. To reach this level requires 13.2 million hectares of irrigated area and 1.1 million tons of fertilizer in excess of the base-period levels.

The third step calculates the production increment from the residual irrigated area (1.6 million hectares) using local varieties, which amounts to 846,000 tons and requires 94,000 tons of fertilizer.

The fourth step computes the fertilizer necessary to bring total production to 108.0 million tons. The necessary output increment is 9.1 million tons and assuming a

[18] As was noted in the discussion of the input basis for 1967-68, this assumption provides a conservative element to the model.

13

Table 7.--India: Model and input base for projecting 1970-71 food grain production at 108 million tons[1]

Inputs and output	Unit	1959-60 to 1961-62 average[2]	Estimates for 1970-71[3]	Estimates of grain output increases from 1959-60 to 1961-62 (Ave.) to 1970-71 with following input increases[4]				
				Area	High yield varieties	Irrigation with local varieties	Fertilizer with non-irrigated local varieties	Total increases
		(1)	(2)	(3)	(4)	(5)	(6)	(7)
Inputs:								
Grain area............	1,000 hectares	116,212	121,000	4,788	0	0	0	4,788
Gross irrigated grain area...........	1,000 hectares	22,318	38,000	915	13,200	1,567	0	15,682
Fertilizer for grain................	1,000 tons	131	2,691	5	1,056	94	1,405	2,560
High-yielding varieties............	1,000 hectares	0	13,200	0	13,200	0	0	13,200
Output:								
Increments...........	1,000 tons	--	27,535	3,299	14,256	846	9,134	27,535
Total................	1,000 tons	80,465	108,000	--	--	--	--	--

[1] The 108 million tons is the level output must reach to attain an annual growth rate of 5 percent from the 1967-68 estimate in table 6.
[2] Irrigated grain area accounts for about 80 percent of total irrigated area. It is assumed that 40 percent of total fertilizer was applied to grain.
[3] Irrigated grain area accounts for about 80 percent of total irrigated area. It is assumed that 75 percent of total fertilizer was applied to grain.
[4] Increases in grain output estimated as follows:
  Column 3 - Yield held constant; production, irrigated area, and fertilizer increases at rate of area increase (3.0 percent). Area taken as trend extrapolation as projected by Holst (20).
  Column 4 - High-yield varieties grown on irrigated land and fertilized at 80 kg./ha.; assumed yield response at 13.5 kg. grain for 1 kg. of fertilizer.
  Column 5 - Fertilizer applied at 60 kg./ha.; assumed yield response of 9 kg. grain for 1 kg. fertilizer.
  Column 6 - Assumed yield response of 6.5 kg. grain for 1 kg. fertilizer and then computed the amount of fertilizer necessary to produce 9.1 million tons of grain--the quantity needed to reach a total output of 108.0 million tons of grain.

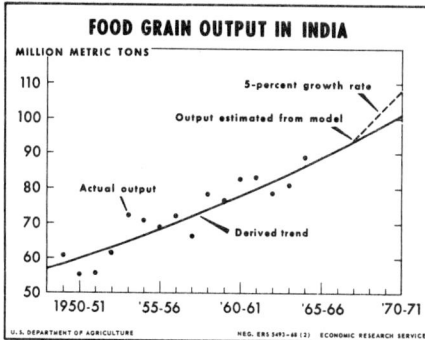

Figure 1

response coefficient of 6.5, the fertilizer requirement is 1.4 million tons.

Therefore, 2.7 million tons of fertilizer together with the gross food grain area of 121 million hectares, the high-yielding variety area of 13.2 million hectares, and an irrigated area of 38.0 million hectares would result in a total output of 108 million tons. The 2.7 million tons of fertilizer represents only that portion of the total supply that is applied to grains.[19] The total fertilizer supply in this case would equal about 3.6 million tons.

Thus, with average weather, 1970-71 grain production should reach 106 million tons and is projected at 108 million tons. The difference between the 108 million and the 106 million set as the objective should be regarded as a safety margin for uncertainties of weather, input supplies and distribution, and assumed response coefficients (fig. 1).

The results of this combination of inputs are somewhat surprising in view of India's fourth plan targets. The gross irrigated area and fertilizer consumption are below the target by about 5 percent and 13 percent, respectively. However, the targets are aimed at the production of 120 million tons of grains, and not the 108 million tons projected here.

But these differences pose the question, "what would be the level of grain output if the targets were fulfilled?" Using the same

---

[19] It has been assumed that 75 percent of the total supply of commercial fertilizers is applied to food grains.

framework as above with the following inputs:

...total grain area	121.0 million hectares
...gross irrigated grain area	40.0 million hectares
...high-yielding varieties area	13.2 million hectares
...fertilizers used for grains	3.1 million tons

the production of grains would total 111 million tons.

On balance therefore, it appears that the objective of an annual growth of 5 percent is attainable with likely supplies of inputs, and could, in fact, be exceeded. But to do so will require a continuous push to effectively acquire and distribute the necessary inputs for cultivator use. Embedded deeply within the framework of the model is the assumption that the growth of India's "agro-industry" will be adequate to serve the rising demands of agriculture. This avoids a host of problems which inevitably will arise during the course of the next 3 years. The scope of this report precludes a comprehensive discussion of these problems but they are important enough to warrant the comments in the following sections.

POLICIES AND PROGRAMS

The preceding section indicates that a 5-percent growth rate in food grain production is technically and economically feasible for the period 1967-68 to 1970-71. Moreover, important foundations for moving out along, or above, this growth line have already been laid and the Center is moving forward to insure such growth.

Previous pessimism about India's grain prospects has been based on two conditions--
...targets for inputs were inadequate to set off and sustain such growth;
...performance has fallen short in fulfilling these low input targets.

In contrast to this past record, recent conditions have changed:
...input targets have been substantially raised; and
...performance against even these higher targets promises to more closely match requirements for their fulfillment.

The Center is pressing vigorously to meet input needs through rapidly expanding domestic production and committing scarce foreign exchange for imports of needed inputs that cannot be supplied domestically. Despite a generally tight budget situation, the Center has greatly increased allocations for agriculture.

High-Yielding Varieties

A dramatic example of the vigor of the Center's efforts to improve agriculture is the importation of Mexican dwarf wheat in 1966. Based upon the results of variety tests in the spring of 1966, the Minister of Food and Agriculture and the State Chief Ministers proposed the importation of $5 million worth of Mexican seed wheat for the 1966-67 rabi (spring) planting. This was cleared through the Finance Ministry within 24 hours. Within a week, Indian seed specialists were in Mexico making field purchases of wheat. The result was that the world's largest seed shipment on record, 18,000 tons, arrived in India within 3 months, in time for planting an estimated 600,000 acres (243,000 hectares).

As mentioned earlier, supplies of high-yielding varieties of rice, wheat, maize, jowar, and bajra are now adequate to plant 15 million acres (6.1 million hectares) in 1967-68 (tables 6 and 15).

Supplying seed for expanding the area of high-yielding varieties to 32 million acres (13.2 million hectares) by 1970-71 should pose no serious difficulty. Basic plant germ plasms from which to develop new varieties with larger yield potentials and improved quality are now available for all major cereal crops. Supplies of such materials for pulse crops are also being collected by USDA geneticists working in cooperation with Indian research agencies under an AID-USDA Participating Agency Services Agreement.

The limited number of trained personnel constitutes a major bottleneck on the speed with which supplies of hybrid jowar, bajra, and maize seed can be increased and therefore affects adversely the rate at which the area of high-yielding varieties can be increased.

In the past, it has often been difficult to maintain high standards of purity and quality of seed supplies--even in some cases for State seed farms. Programs to insure purity and quality of commercial seed stock need to be strengthened throughout most of India. A step in this direction was the recent passage of a National Seed Law to provide quality controls through seed certification and registration procedures. Implementing legislation by the States, which is now under discussion, will be necessary to make the National Seed Law effective.

In the multiplication of improved varieties, heavy emphasis has heretofore been placed on State seed farms. Currently, however, the private sector is being used extensively to supplement State seed farms, which will help to insure adequacy of seed supplies needed to sustain a rapid rate of growth. It is not clear, however, that much encouragement is being given to use of private firms to produce seed.

Fertilizers

There has been a spectacular change in the fertilizer situation during the past 2 years. Previously there was concern that supplies would exceed demand and attention had been focused on avoiding a possible glut. But with the recent technological developments and relatively high food grain prices, present efforts are directed to meeting a rapidly increasing demand for fertilizers. This shift is demonstrated in various ways:

...Fertilizer availability targets for the fourth plan are up 4 to 5 times over third plan availabilities; domestic production targets show the greater increase but foreign exchange has been committed to imports necessary to meet the balance of targets.

...India's performance in the first two crop years of the fourth plan (1966-67 and 1967-68) has been creditable. Nitrogen available for the first agricultural year of the plan was over 900,000 metric tons--an increase of 55 percent over the previous year and about 90 percent of the goal. Similarly, availability of nitrogen for the second agricultural year will increase 45 percent to over 1.3 million metric tons. Availability of P_2O_5 doubled the first year and increased an additional 50 percent in the second; K_2O availability showed smaller but still significant gains.

...There have been N and P_2O_5 shortfalls in the production sector, stemming from shortages of raw materials and drought-aggravated power shortages. Nonetheless, production has risen substantially, both absolutely and as a percentage of target fulfillment. Even more encouraging has been the Center's evident willingness to commit very scarce foreign exchange and to carry through on importations of N and P in excess of import targets as well as improving the imports of K.

...Earlier commitment of funds against pending budgets has permitted more timely placing of fertilizer orders in the last 2 years.

...Difficulties experienced by the State Trading Corporation in obtaining sulfur in January 1967 led to formation of a joint Government-Industry Fertilizer Allocation Committee to review import requirements and prospective contracts. Current estimates indicate that the 600,000-ton annual requirement will be met and possibly exceeded. Proliferation of buyers, including private traders, and the freedom to develop a variety of contract patterns have widened the supply prospects and resulted in price benefits on longer term contracts.

...Contract negotiations to build manufacturing plants have been expedited.

Irrigation

Compared with the third plan, the fourth has given emphasis to minor irrigation expansion;[20] allocations for minor irrigation increased by 93 percent while those for major and medium increased only 47 percent--a good part of which represents completion of previous projects.

These target increases should also be viewed in the context of third plan performance, which exceed targets for minor irrigation projects but fell short of major to medium targets. For the first year of the current plan period, 28 percent of the minor irrigation target area was covered. There has also been a significant shift in the pattern of minor irrigation programs. In the first plan, the additional areas irrigated by surface (tanks[21] and canals) and ground (wells) water development were about equal, whereas in the fourth plan, the area increment expected from ground water development is more than double that from surface water.

Of the various types of irrigation wells to be developed, expansion of well construction programs are clearly emphasizing private over public ownership. Compared with the previous plan, the number of additional private tubewells is planned to increase nearly 160 percent while public tubewells will increase 100 percent; the former will serve an area nearly twice as great as the latter.

The planned increase in motorized pumps for wells (243 percent of third plan achievement for electric and 112 percent for diesel) will further reinforce the production potential from the increased well construction in the fourth plan. For example, the State of Uttar Pradesh originally planned to install 10,000 pumps in 1966-67; later, with drought conditions prevalent, the target was raised to 17,000 and was reached before the end of the fiscal year. Rural electrification has a high priority in the current plan. The Center estimates, as a result, that they will be able to remove the present 2-year delay in well installation within the next few years. This development would obviate the alleged preference given to public wells in obtaining power connections in some areas. It is estimated that the rate of well construction increased 50 percent between 1965-66 and 1966-67; further increases are expected in 1967-68.

A variety of measures are being taken to increase the effectiveness of irrigation programs. The Ayacut (command area) Development Program was recently organized at the Center to promote integrated local development of irrigation projects in such related spheres as shaping of channels, changing cultivation practices, assuring needed inputs, and water management measures. More generally, a Water Utilization Unit has been organized within the Ministry of Food and Agriculture to direct the Ayacut Program and to promote better utilization of water resources through coordination of irrigation agencies. Through the Ayacut Program and the Water Utilization Unit, there should be gains in integrated local focus as well as better top-level coordination of irrigation activities.

There has been an appreciable increase in credit resources through established institutions (Land Development Banks, and Agricultural Refinance Corporation) for financing wells and grading land. The formation of new credit institutions for similar purposes is now under consideration.

Plant Protection Materials

The advent of the high-yielding varieties highlights the need for more disease and pest control measures. The new varieties

---

[20] Irrigation projects in India are classified according to cost: major ($6.7 million plus); medium ($0.13 to $6.7 million); and minor (less than $133,300).

[21] Ponds, lakes, or reservoirs are commonly referred to in India as "tanks," and driven wells as "tubewells."

are amenable to much denser planting; the larger plant populations lead directly to greater insect populations, and provide an environment for the spread of disease. With traditional varieties, the profitability of plant protection measures was marginal at best, but a comprehensive control program is profitable for the high-yielding varieties.

Plant protection benefited from the Center import liberalization in 1966 which freed the importation of needed technical ingredients; production of plant protection materials for 1967-68 is estimated to be nearly 20 percent greater than for the preceding year. The Center has recently agreed to continue subsidizing the cost of producing pest control materials.

The area covered by pest control measures has increased sharply from 16.6 million hectares in 1965-66 to 25.5 million hectares in 1966-67; 51 million hectares are planned for 1967-68. This increase, however, does not indicate the effectiveness of such action. The area may or may not be thoroughly covered; the actual need for protection--from a locust infestation, for instance--may vary greatly from year to year; climatic variations also influence the need for protection; and there are many alternative means for protection as well as alternative protection needs. However, a "survey and warning" system is being established to arrest any potentially serious infestation before epidemic proportions are reached.

Transport Facilities

To achieve the annual growth rate of 5 percent in food grain production will require even higher rates of growth for all major inputs except land. The projected annual rates of growth are 1.0 percent for grain area; 5.9 percent for irrigated grain area; 29.5 percent for the area under high-yield varieties; and 19.5 percent for fertilizer consumption (table 14).

These high rates must be accompanied by a substantial expansion in the facilities that supply and distribute farm inputs to the cultivator. In fact, the 5-percent annual growth rate in grain production in itself will require additional marketing facilities that can effectively transfer the food grains from the producer to the consumer.

Transportation is the underpinning of an agricultural marketing and distribution system. In almost every developing country, the network of access roads between farms and local market towns is still inadequate. In India, there is only about 0.7 mile of road per square mile of cultivated land, compared with about 4 miles in the United Kingdom, France, Japan, and the United States.

It has been estimated that in India a million miles of roads will have to be constructed to satisfy the access needs of 580,000 villages throughout the country. Only 11 percent of these villages now have reasonably adequate roads and one out of three is more than 5 miles from a satisfactory road ([20]).

The most important transport program for Indian economic development in the fourth plan would be to concentrate on the agricultural sector to permit the distribution of necessary farm supplies and to make possible the marketing of farm commodities. With sharply rising supplies of farm inputs and the increased output that is anticipated from these inputs there is an immediate urgency in developing an adequate transport network.[22]

Agricultural Credit

In the past year there has been direction in forming new credit institutions (or reorganizing existing institutions) and in increasing funds for credit purposes, including:

> ...For 1967-68 expansion of credit funds for agricultural purposes, the Center has published commitments to expand credit by over Rs. 95 crores[23] ($126.7 million) with at least an additional Rs. 5 crores ($6.7 million) promised if performance by credit institutions in lending is adequate: Nearly Rs. 17 crores ($22.7 million) are allocated to medium/long term facilities (1.4 to Land Development Banks and 15.5 to the Agricultural Refinance Corporation), an additional Rs. 9 crores ($12.0 million) to medium-term lending (the newly formed Agro-Industries Corporations) and Rs. 70 crores ($93.3 million) to short-term lending (25 through the cooperatives and 45 in support of input program lending). The additional 5 crores ($6.7 million) promised will go to the Land Development Banks upon demonstration of effectiveness of the new levels.

---

[22] For a discussion of a suggested transportation program for India see ([20]) vol. II, pp. 589-592.

[23] The rupee (Rs.) is the basic monetary unit in India. Since June 6, 1966, it has been officially valued at $0.1333 (U.S.). A crore is 10 million.

...The recent creation of the Agro-Industry Corporations and pending Government legislation setting up Agricultural Development Corporations in States having weak cooperative lending institutions.

...The Center has been considering suggestions for still other agricultural credit institutions or patterns of rural lending, especially those related to fertilizer distribution and construction of wells.

...Recently, the Association of Indian Commercial Banks has announced the intention of setting aside a fund of Rs. 350 crores ($466.7 million) for agricultural production lending. This step was taken as a partial answer to the growing public criticism of the unwillingness of commercial banks to share the responsibility for rural credit needs. While the details have not yet been worked out on the operation of this fund, there are indications that it will be directed first toward greater credit facilities for individual cultivators and then for utilization by input suppliers and the distribution channels.

Agricultural Research and Education

In the field of research, the Rockefeller Foundation, through its coordinated research programs for hybrids and new wheat and rice varieties, has contributed substantially to the present promise of the high-yielding varieties program. These efforts are being augmented by the research programs conducted by the Indian Agricultural Research Institute at the Center and various research facilities in the States. A recently signed agreement between the Center and the International Rice Research Institute is a further indication of future research emphasis for this important food crop.

Agricultural research of late has been coming closer to field operations: In 1967-68 scientists of the Indian Council of Agricultural Research (ICAR) will continue to organize national demonstration projects in the field which will be supplemented in several States by demonstration farms with the assistance of an agricultural extension staff. A coordinated research program for about 20 commodities in various States has been undertaken by the ICAR in collaboration with the State Governments.

The AID Mission programming of Field Problems Research Teams is a healthy development relating research, extension, and operations. Currently operating in four States, these five-man teams are actively engaged in promoting better use of fertilizer, seeds, plant protection inputs, and water management by expediting and promoting the linkage between field experiences, research facilities, and extension activities within the States. Working with State agencies on the one hand and agricultural universities on the other, these field units will also underscore the work of the Mission's agricultural universities program which is oriented to a more pragmatic and unified relation between teaching, research, and extension.

The degree of success experienced by the Center in developing the foregoing and related programs will determine the long-run ability of the agricultural sector to maintain the projected growth trend.

Incentives

The situation with respect to price policies is currently more uncertain and confused than that relating to any other major requirement for sustaining a rapid rate of growth in food grain production. Creation of the Agricultural Prices Commission in 1965 indicates an awareness of the need for more rational price policies. Whether actual improvements have been made in India's agricultural price policies remains to be seen.

Prices of food grains at the time this report was written were favorable throughout India, a fact best attested to by the current demand for fertilizer and other inputs. Price relationships among States and between commodities are, however, greatly distorted and are wholly inconsistent with objectives of efficiency in allocation of scarce inputs and with that of efficiency in food distribution (tables 12 and 13). The reason for this is the existence of the State zonal system prohibiting free interstate trade in grains.

India's zonal system is currently depressing prices of grains in localities having the largest comparative advantage in their production and inflating grain prices in deficit producing areas. Under present demand-supply relationships applicable to fertilizers and other major inputs, these distorted price relationship have little effect upon the overall amount of these inputs now being used. However, unless counteracted by appropriate administrative allocative controls, such distortion of price relationships must be an added source of inefficiency in the allocation of scare inputs that are strategic to India's food needs. There will inevitably be inefficiency in allocation of the nation's supplies of seeds, fertilizers, and other inputs simply because of the speed with which these supplies have been increased. This "administrative" inefficiency is an added waste at a time when efficiency is of the utmost importance, not only for achieving the nation's food production targets, but for the conservation of foreign exchange.

Currently, India has a system of support prices, but the announced level of these supports falls so far below both current price levels and those for 1962-63 to 1964-65 that they can hardly be called incentives. As India's food grain production approaches a 5-percent per year growth rate, it will likely cause a downturn in food grain prices from their presently high scarcity levels. This in itself would pose a very delicate and difficult analytical problem which could be the next hurdle for Indian administrators to cope with: How to determine the level of price supports needed to insure adequate producer incentives without, however, distorting price relationships and constraining the role of free market prices.

REFERENCES

(1) Abel, Martin E. and Rojko, Anthony S.
  1967. World Food Situation. U.S. Dept. Agr., Foreign Agr. Econ. Rpt. 35, Aug.
(2) Central Statistical Organization, Cabinet Secretariat, Government of India
  Estimates of National Income, 1950-51 and 1955-56 to 1959-60.
(3) Directorate of Economics and Statistics, Ministry of Food, Agriculture, Community Development and Cooperation, Government of India
(4) _____
  1961-67 Agricultural Prices in India, Various Issues.
(5) _____
  1961-67 Agricultural Situation in India, Various Issues.
(6) _____
  1961-67 Bulletin on Food Statistics, Various Issues.
(7) _____
  1966. Growth Rates in Agriculture, 1949-50 to 1964-65, March.
(8) _____
  Land Utilization Statistics--All India, 1950-51 to 1962-63.
(9) _____
  1967. Area, Production and Yield of Principal Crops in India, 1949-50 to 1966-67; Summary tables. Oct.
(10) Fertilizer Association of India
  1967. Fertilizer Statistics.
(11) Ford Foundation
  1967. A Richer Harvest, Oct.
(12) Foreign Regional Analysis Division, Economic Research Service
  1964. The Word Food Budget, 1970. U.S. Dept. Agr., Foreign Agr. Econ. Rpt. 19, Oct.
(13) Hall, William F.
  1964. Agriculture in India. U.S. Dept. Agr., Econ. Res. Serv., ERS-Foreign 64, Jan.
(14) Indian Council of Agricultural Research and Cooperating Agencies
  1966. Progress Report of the All India Coordinated Rice Improvement Project, Kharif, 1966.
(15) _____
  1967. Progress Report of the Coordinated Millets Improvement Program, 1965-66.
(16) Institute of Agricultural Research Statistics
  1964. Yardsticks of Additional Production of Certain Food Grains, Commercial, and Oilseed Crops. New Delhi.
(17) International Bank for Reconstruction and Development
  1967. Indian Economic Policy and the Fourth Five-Year Plan. Vol. II. Agricultural Policy in India. March.
(18) Leftwich, Richard H.
  1961. The Price System and Resource Allocation. Holt Rinehart and Winston, New York.
(19) Lewis, John P.
  1964. Quiet Crisis in India, Doubleday, Garden City, New York.
(20) President's Science Advisory Committee
  1967. The World Food Problem. Volumes I and II. May.
(21) Sen, S. R.
  1967. Growth and Stability in Indian Agriculture. Agricultural Situation in India. Jan.

APPENDIX

Comments on a Report of the President's Science Advisory Committee, The World Food Problem (20)

General Report.--There are no basic inconsistencies between the report of the President's Science Advisory Committee (PSAC), The World Food Problem, and the analysis presented in this report. The former is properly global in its view. It is addressed to a very wide range of problems treated in broad general terms without assignment of priorities and without reference to specified constraints in respect to budgetary considerations, input availabilities, and many other items that are specific to our own situation. In developing this report, we have attempted to assess development potentials and requirements under conditions that are specific to India. We have attempted to project a program that we believe is attainable, yet challenging, within limits of budgetary, resource, organizational, and other constraints applicable to India.

In our analysis we have also placed heavy emphasis upon programs with good promise of large increases in food production in the near term. India's current food crisis, very recent but large improvements in food grain technology, and recent shifts of emphasis in the Indian Government's food production policies all make this emphasis upon achieving large early increases in output desirable. Measures to achieve these shortrun gains will, however, help to strengthen long-term development programs, including those of agricultural education, extension, and research.

This report, as well as recent policy emphasis of both the AID Mission to India and the Government of India is fully consistent with the high priority assigned in the PSAC report "to providing production inputs essential to accelerating agricultural productivity."

The Mission's program in support of agricultural education, extension, and research is being strengthened by the addition of U.S. agricultural experts to work jointly with Indian Universities and State Departments of Agriculture in production promotion activities.

The Holst Paper.--Compared with the "self-sufficiency" figure of 113.5 million tons of food-grains needed for 1971 in the Holst model (a chapter in (20, vol II), our figure (106 million tons) of that which is attainable is conservative. However, there are several differences between the information and assumptions used here and those used by Holst:

...His model includes in the concept of self-sufficiency an increase in the nutritional level of the population which would increase the total needed by some unspecified amount.

...Drawing on seed and fertilizer responses derived from 1963-64 data, he projects from a technological base which has been dramatically altered by the unexpected and rapidly spreading introduction of new varieties of wheat, paddy, and hybrids. These high-yielding varieties, when coupled with the equally rapid and dramatic rise in fertilizer availability, will produce in the immediate future, and on a sustained basis thereafter, levels of production not anticipated until years later in his model.

...Another point of difference is the historical growth rate of agriculture and, therefore, the normative base from which he projects. We have demonstrated that a good part of the flattening of the growth curve which Holst notes in the late 1950's and early 1960's can be attributed to markedly poor weather. Grain prices were relatively low which would have also contributed to the flattening of the curve, but prices have shifted greatly in favor of grains since 1963-64. Therefore, given the higher base level for projections which we feel is justified and in view of the input/output changes consequent on the new technology now well in process in India, our estimates can be viewed as more conservative than Holst's.

...Finally, Holst uses a loss figure which is much larger than that customarily used by either the AID Mission or the Center. While it may reasonably be argued that some higher loss figure is justified, currently there is no firm basis for making it as high as in the Holst model nor does his projection appear to consider the determined efforts now being made by the Center to improve storage facilities; to rapidly expand the plant protection program; to develop improved grain varieties;

and to initiate rodent control programs. All of these efforts are having, and will continue to have, an influence in reducing losses. (Loss estimates are not relevant to the output projections made in this analysis, but they do bear on the extent to which these output levels fulfill the objective of self-sufficiency in food grain consumption.)

Model for Estimating Food Grain Output

A simple model is constructed in this study to measure the marginal product of food grains resulting from increases in basic agricultural inputs. The model is discussed subjectively in the text. A more formal presentation is given below:

$O' = O_b + M.P.$, when

O' = food grain output estimate
O_b = food grain output in base period (1959-60 to 1961-62)
M.P. = Marginal product of food grain resulting from increases in basic agricultural inputs where

$M.P. = O_a + O_{hyv} + O_i + O_f$ when

O_a = food grain output resulting mainly from area expansion.
O_{hyv} = food grain output resulting from increased use of high yielding varieties, given irrigation and fertilization at the rate of 60 kilograms per hectare.
O_i = food grain output resulting from residual increase in irrigated food grain area, assuming area planted with local varieties and fertilized at the rate of 40 kilograms per hectare
O_f = food grain output resulting from residual increase in fertilization, assuming it is applied to local varieties and

$O_a = (\Delta A)(O_b)$ when

ΔA = percentage change in food grain area

$O_{hyv} = (13.5)(60.0)(\Delta HYV)$ when

13.5 = assumed response ratio where one unit of fertilizer yields 13.5 units of grain
60.0 = rate of fertilization in kilograms per hectare
ΔHYV = change in area of high yielding food grain varieties in thousand of hectares

$O_i = (9.0)(40.0)[\Delta I - (\Delta A \cdot I_b + \Delta HYV)]$ when

9.0 = assumed response ratio where one unit of fertilizer yields 9.0 units of grain
40.0 = rate of fertilization in kilograms per hectare
ΔI = change in irrigated food grain area in thousand hectares
I_b = irrigated food grain area in the base period in thousands of hectares

$O_f = (6.5)\left[\Delta F - (A \cdot F_b + O_{hyv}/13.5 + O_i/9.0)\right]$

6.5 = assumed response ratio where one unit of fertilizer yields 6.5 units of grain
ΔF = change in fertilizer consumption in thousand tons
F_b = fertilizer consumption in base period in thousand tons

Given the following for 1967-68:
O_b = 80,465,000 tons
ΔA = 1.1 percent
ΔHYV = 6,100,000 hectares
ΔI = 9,682,000 hectares
I_b = 22,318,000 hectares
ΔF = 1,444,000 tons
F_b = 131,000 tons

Solve for O'_{67-68}
O_a = (.011)(80,465)
     = <u>885,000</u> tons

O_{hyv} = (13.5)(60.0)(6,100)
         = <u>4,941,000</u> tons

O_i = (9.0)(40.0)[9,682 - (245 + 6,100)]
     = (360.0)(3,337)
     = <u>1,197,000</u> tons

O_f = (6.5)[1,444 - (1 + 366 + 133)]
     = (6.5)(944)
     = <u>6,136,000</u> metric tons and

M.P. = 885 + 4,941 + 1,197 + 6,136
     = 13,159,000 tons and

O'_{67-68} = 80,465 + 13,159
            = <u>93,624,000</u> tons

TABLES

Table 8.--India: Food grain production by States, 1964-65 to 1966-67

State	1964-65[1]	1965-66[1]	1966-67[2]	(3/2)	(3/1)
	(1)	(2)	(3)	(4)	(5)
	------Thousand metric tons------			------Percent------	
Northeast:					
Assam............................	1,966	1,903	1,848	97	94
Bihar............................	7,532	7,148	4,225	59	56
West Bengal......................	6,260	5,448	5,394	99	86
Orissa...........................	4,946	3,737	4,246	114	86
Nagaland.........................	43	43	47	109	4
North and Northwest:					
Uttar Pradesh....................	15,289	13,311	12,459	94	81
Punjab...........................	[3] 7,224	3,453	4,179	121	58
Rajasthan........................	5,308	3,839	4,338	113	82
Jammu & Kashmir..................	566	480	648	135	114
Haryana..........................	([4])	1,977	2,606	132	([4])
Central and West Central:					
Madhya Pradesh...................	10,209	6,807	6,347	93	62
Gujarat..........................	2,816	2,305	2,310	100	82
Maharashtra......................	6,838	4,722	6,216	132	91
South:					
Andhra Pradesh...................	7,634	6,219	7,660	123	100
Madras...........................	5,739	5,251	5,830	111	102
Mysore...........................	4,531	3,134	4,077	130	90
Kerala...........................	1,150	1,025	1,123	110	98
Union Territories:	948	1,228	1,497	125	158
Total all India................	88,996	72,030	75,049	104	84
Total minus Bihar............	81,464	64,882	70,824	109	87
Total minus Bihar, U.P. & M.P.................	55,966	44,764	52,018	116	93

[1] Partially revised estimates.
[2] Final estimates.
[3] Includes Haryana.
[4] Included under Punjab.

Source: (3).

Table 9.--India: Food grain production by crops, 1964-65 to 1966-67

	1964-65[1]	1965-66[1]	1966-67[2]
Cereals:			
Kharif[3]	--------Million metric tons--------		
Rice:			
Autumn	16.15	11.90	13.34
Winter	21.53	17.61	15.36
Total rice	37.68	29.51	28.70
Jowar	6.26	4.78	5.09
Bajra	4.46	3.65	4.50
Maize	4.66	4.76	4.99
Ragi	1.90	1.18	1.60
Small millets	1.95	1.65	1.67
Total kharif cereals	56.91	45.53	46.55
Rabi[4]			
Rice: Summer	1.35	1.15	1.74
Wheat	12.29	10.42	11.53
Barley	2.52	2.38	2.45
Jowar	3.49	2.75	3.86
Total rabi cereals	19.65	16.70	19.58
Total cereals	76.56	62.23	66.13
Pulses			
Kharif	3.61	3.09	3.07
Rabi	8.83	6.71	5.85
Total pulses	12.44	9.80	8.92
Total food grains	89.00	72.03	75.05

[1] Partially revised estimates.
[2] Final estimates.
[3] Kharif refers to the fall and winter harvest.
[4] Rabi refers to the spring harvest.

Source: (3)

Table 10.-- India: Yield of major food grain crops shown with and without irrigation, 1964-65

Crops	Irrigated		Nonirrigated		Yield difference (2)-(4) Kg/Ha.
	Hectares (000)	Yield (Kg/Ha.)	Hectares (000)	Yield Kg/Ha.	
	(1)	(2)	(3)	(4)	(5)
Rice............................	13,424	1,371	22,940	899	472
Wheat...........................	4,858	1,173	9,602	766	407
Jowar...........................	681	734	17,257	536	198
Bajra...........................	268	560	11,458	378	182
Maize...........................	551	1,452	4,067	949	503
Ragi............................	347	1,009	2,090	741	268
Barley..........................	1,294	1,159	1,390	736	423
Gram............................	1,374	873	7,522	610	263
Other...........................	966	621	18,444	434	187
Total food grains..............	23,763	1,229	93,779	836	393

[1] These yield differences reflect not only the influence of irrigation on yields but that also of associated differences in inputs of fertilizers, seeds, pesticides, and management. It is believed that most of the fertilizers used in India in 1964-65 was used on irrigated crops; also that improved seeds are more commonly used on irrigated than on nonirrigated land.

Table 11.--India: Statewide growth rates (compound) of agricultural production, area and productivity, 1952-53 to 1964-65

State	Production	Area	Productivity
	----------Percent----------		
Above average:			
Punjab......................	4.56	1.90	2.61
Gujarat.....................	4.55	0.45	4.09
Madras......................	4.17	1.10	3.04
Mysore......................	3.54	0.81	2.71
Himachal Pradesh............	3.39	0.71	2.67
Fair:			
Bihar.......................	2.97	0.71	2.25
Maharashtra.................	2.93	0.44	2.45
Rajasthan...................	2.74	2.85	- 0.11
Andhra Pradesh..............	2.71	0.26	2.45
Madhya Pradesh..............	2.49	1.28	1.21
Orissa......................	2.48	0.81	1.66
Low:			
Kerala......................	2.27	1.30	0.96
West Bengal.................	1.94	0.59	1.34
Uttar Pradesh...............	1.66	0.72	0.94
Assam.......................	1.17	1.25	- 0.08
All India................	3.01	1.21	1.77

Source: (7).

Table 12.--India: Annual average wholesale price for rice, 1961-66

State	Variety	No. of markets	1961	1962	1963	1964	1965	1966
					Rupees per quintal			
Andhra Pradesh	Akkulu	(3)	55.62	54.99	54.39	61.24	[1] 63.07	[1] 65.02
Assam	Sali	(3)	51.12	55.65	59.92	66.06	[1] 65.58	[1] 65.14
Bihar	Coarse	(5)	55.73	57.74	63.54	70.51	85.08	126.43
Kerala	Coarse	(2)	60.91	58.57	60.90	71.20	[1] 63.50	[1] 68.67
Madhya Pradesh	Coarse	(3)	41.52	43.92	52.26	58.13	[1] 58.23	[1] 64.80
Madras	Medium	(3)	60.24	59.05	57.21	65.33	[1] 66.03	[1] 65.10
Maharashtra	Coarse	(3)	55.78	52.20	59.74	68.92	[1] 70.05	[1] 69.72
Mysore	Coarse	(3)	59.44	59.59	53.53	66.80	89.36	116.60
Orissa	Coarse	(4)	39.71	48.86	61.59	61.20	[1] 59.90	76.56
Punjab	Coarse	(1)	44.21	44.21	44.21	50.17	[1] 60.00	[1] 60.00
Uttar Pradesh	Coarse	(3)	51.51	52.20	54.34	69.16	[1] 65.67	129.09
West Bengal	Common	(5)	52.77	61.26	77.73	64.05	66.11	[1] 72.00

[1] Statutory controlled prices fixed by State governments (average).

Source: (4), (5), and (6).

Table 13.--India: Annual average wholesale price for wheat, 1961-66

State	Variety	No. of markets	1961	1962	1963	1964	1965	1966
			----------	----------	Rupees per	quintal ----	----------	----------
Bihar	White	(1)	49.59	48.92	49.96	69.30	97.00	107.50
Gujarat	Red	(1)	51.20	51.02	51.17	62.98	68.99	72.98
Madhya Pradesh	White	(3)	36.62	40.28	40.48	56.55	60.57	60.16[1]
Punjab	Coarse	(1)	39.41	42.49	40.34	51.47	58.40	70.88
Rajasthan	Coarse	(1)	43.72	41.69	39.38	52.32	51.23	74.14
Uttar Pradesh	Red	(1)	38.78	36.09	39.15	65.15	75.21	69.81
	White	(2)	40.09	39.13	41.68	71.00	79.68	78.91
	Dara	(1)	41.86	40.70	42.85	72.08	61.67	79.17

[1] Statutory prices fixed by the State governments (average).

Source: (4), (5), and (6).

Table 14.--India: Projected annual inputs required to achieve an annual compound growth rate in food grain output of 5 percent from 1967-68 to 1970-71[1]

Inputs and production	Unit	1967-68	1968-69	1969-70	1970-71	Annual compound rate of increase 1967-68 to 1970-71
Food grain area	1,000 ha.	117,500	118,675	119,885	121,000	1.0
Gross irrigated food grain area	1,000 ha.	32,000	33,890	35,890	38,000	5.9
High-yield varieties	1,000 ha.	6,100	7,900	10,200	13,200	29.5
Fertilizers	1,000 tons	1,575	1,880	2,250	2,691	19.5
Food grain production	1,000 tons	93,624	98,212	103,024	108,000	5.0

[1] The inputs and production for 1968-69 and 1969-70 are interpolations between 1967-68 and 1970-71 (see tables 6 and 7). They should be considered as only general trends to achieve the desired 5-percent annual growth objective discussed in this report.

Table 15.--India: High-yielding varieties program--revised targets for 1967-68 (kharif and rabi/summer)

State	Paddy		Maize		Jowar		Bajra		Wheat		Total	
	Kharif	Rabi	Kharif	Rabi	Kharif	Rabi	Kharif	Rabi	Kharif	Rabi	Kharif	Rabi
	------Thousand acres------											
Andhra Pradesh...	700	720	65	30	70	116	70	20	--	--	905	886
Assam............	71	7	13	2	--	--	--	--	--	2	84	11
Bihar............	500	¹ 300	200	¹ 220	--	--	--	--	--	¹ 500	700	¹ 1,020
Gujarat..........	160	--	50	8	6	--	300	100	--	314	516	422
Haryana..........	23	--	10	--	--	--	30	--	--	200	63	200
Jammu and Kashmir	100	--	30	--	--	--	10	--	--	20	140	20
Kerala...........	250	500	3	--	--	--	--	--	--	--	253	500
Madhya Pradesh...	50	--	100	¹ 1	95	¹ 10	16	--	--	120	261	¹ 131
Madras...........	800	100	1	9	7	143	21	28	--	--	829	280
Maharashtra......	400	100	150	150	1000	800	300	--	--	200	1850	¹ 1,250
Mysore...........	200	¹ 70	50	¹ 45	250	¹ 90	50	1.50	--	¹ 10	550	¹ 216.5
Orissa...........	220	140	12	8	3	0.1	--	--	--	5	235	153.1
Punjab...........	50	--	100	--	--	--	100	--	--	1,000	250	1000
Rajasthan........	2	--	45	--	10	--	80	--	--	125	137	125
Uttar Pradesh....	250	--	325	--	20	--	80	--	--	2,000	675	2,000
West Bengal......	300	75	10	5	--	--	--	--	--	40	310	120
Himachal Pradesh.	20	--	17	--	--	--	--	--	--	20	37	20
Delhi............	¹ 0.5	--	¹ 1	¹ 1	--	--	¹ 20	--	--	¹ 5.75	¹ 21.5	¹ 5.75
Goa..............	¹ 25	¹ 5	¹ 0.6	¹ 5	--	¹ 0.2	--	--	--	--	¹ 25.6	¹ 10.20
Pondicherry......	¹ 15	¹ 5	--	--	--	¹ 0.2	--	¹ 0.25	--	--	¹ 15.0	¹ 5.45
Total	4,136.5	2,022	1,182.6	483	1,461	1,159.5	1,077	149.75	--	4,561.75	7,857.1	8,376.00

¹ Provisional.

APRIL 1969

TECHNOLOGICAL CHANGE IN AGRICULTURE

Effects and Implications
for the Developing Nations

FOREIGN AGRICULTURAL SERVICE
U.S. DEPARTMENT OF AGRICULTURE
cooperating with
AGENCY FOR INTERNATIONAL DEVELOPMENT

TECHNOLOGICAL CHANGE IN AGRICULTURE

Effects and Implications for the Developing Nations

by

Dana G. Dalrymple
International Economist

International Development
Foreign Agricultural Service
U. S. Department of Agriculture

in cooperation with
Agency for International Development

Washington, D. C. 20250

PREFACE

The effects of technological changes in agriculture in the less developed nations are a matter of increasing concern. As efforts to expand agricultural production begin to pay off, the many and complex ramifications of technological change become more and more evident.

Untangling the effects of technology is a vast and difficult project. It is one which requires knowledge of many disciplines: economics, political science, anthropology, sociology, and others. It is one which could be the subject of many booklength studies. Yet, curiously, little seems to have been done.

The purpose of this bulletin is to provide an introduction to some of the major effects of technological change in agriculture. Primary attention is given to economic aspects, but an attempt has been made to reflect a broader viewpoint. The report is basically a survey; it is based in part on materials reviewed or prepared for administrative use.

Hopefully, the bulletin will be of value as a reference for those actually engaged in development programs or development planning. In addition, it may provide useful background for development scholars and/or serve as a catalyst for further thought and research. It is, in any case, only a start.

ACKNOWLEDGEMENTS

A number of individuals were of assistance in the preparation of this report. Among these were colleagues in the Department of Agriculture who reviewed earlier drafts of the manuscript.
Drs. Dale Adams of the Agency for International Development and Vernon Ruttan of the University of Minnesota provided helpful references and suggestions.

CONTENTS

I.	**INTRODUCTION**	1
	A. Need for Broader Perspective	1
	B. Interrelations of Technology	2
	C. Terms of Reference	3
II.	**NATURE OF TECHNOLOGICAL CHANGE**	5
	A. Economic Definitions of Technological Change	5
	1. Changes in Production Functions	5
	2. Addition of Production Functions	6
	B. Technological Change Within Agriculture	7
III.	**ADOPTION PROCESS FOR AGRICULTURAL TECHNOLOGY**	10
	A. General Characteristics	10
	B. Why is Technology Adopted?	11
	1. Farm Level Decisions	11
	a. Economic Reasons for Adoption	11
	b. Other Reasons for Adoption	12
	2. Governmental Level Decisions	13
	C. Factors Influencing Rate of Adoption of Technology	14
	1. Characteristics of the Technology	14
	2. Characteristics of the Adopter	15
	3. Characteristics of the Economy	15
	D. Possible Rejection of Technology	16
IV.	**IMPACT OF CHANGES IN AGRICULTURAL TECHNOLOGY**	21
	A. Impact of the Farm Level	21
	1. Economic Implications	21
	a. Primary Effects of Technology	21
	(1) Influence on Supply	21
	(2) Influence on Farm Income	22
	(3) Influence on Employment	25
	b. Secondary Effects of Technology	25
	2. Social/Political Implications	27
	B. Impact at the National Level	27
	1. Economic Effects	27
	2. Social/Political Effects	28
	3. Ecological Effects	29
	C. Impact at the International Level	30
	1. Comparative Advantage and New Products	30
	2. Role of Export Industries	30
	D. Interrelationships	31

V.	HIGH-YIELDING VARIETIES OF GRAIN	35
	A. Background of New Varieties	35
	1. Foundation-Sponsored Research	35
	2. Spread of New Varieties	36
	B. Nature of Technical Changes	36
	1. Functions Involved	36
	2. Need for Technical Base	38
	C. Adoption Process for New Varieties	39
	1. Factors Influencing Rate of Adoption	39
	2. Role of Government in Adoption	40
	3. Permanency of Adoption	41
	D. Impact of the New Varieties	41
	1. Economic Impact	41
	a. Agricultural Output	41
	(1) Quantitative Effect	42
	(2) Qualitative Effect	43
	(3) Influence on Cropping Pattern	43
	b. Financial Returns	44
	(1) Changes in Farm Prices and Costs	44
	(2) Changes in Farm Income	45
	2. Social/Political Impact	45
	3. National and International Implications	47
VI.	MECHANIZATION OF AGRICULTURE	52
	A. Background of Mechanization	52
	1. Development and Spread	52
	2. Power Available	54
	B. Nature of Technical Changes	54
	1. Functions Involved	54
	2. Need for Technical Base	55
	C. Adoption Process for Machinery	56
	1. Size and Nature of Farms Involved	56
	2. Reasons for Adopting	57
	3. Acquisition Process	58
	D. Impact of Mechanization	59
	1. Economic Impact	59
	a. Primary Effects of Mechanization	59
	(1) Production and Costs	59
	(2) Reduced Losses	60
	(3) Farm Organization	60
	b. Secondary Effects of Mechanization	60
	2. Social/Political Impact	61
	3. Balancing Economic and Social Considerations	62

VII.	POLICY IMPLICATIONS OF TECHNOLOGICAL CHANGE		66
	A. Need for Evaluation		66
	1. Benefits of Technology		66
	2. Problems of Technological Change		66
	a. Uneven Distribution of Benefits		66
	b. Disruptions Associated with Change		67
	B. Criteria for Evaluation		68
	C. Policy Issues		69
	1. Factor Combinations and Investments		69
	2. Distribution of Gains and Losses		70
	a. Within Agriculture		70
	b. Between Producers and Consumers		72
	D. Concluding Remarks		72
VIII.	SELECTED BIBLIOGRAPHY		76
	A. Books		76
	B. Bulletins and Reports		77
	C. Articles		77
	D. Unpublished Papers		81

TABLES

1. Effect of Technological Change in Agriculture on Net Farm Income Under Varying Conditions — 24

2. Estimated Area Planted to New High-Yielding Varieties of Wheat and Rice in the Less Developed Nations — 37

3. Estimated Number of Tractors in Use in Major Regions of the World — 53

FIGURES

1. Adoption Process for a New Technology — 10

2. Proportion of Corn Acreage Planted with Hybrid Seed — 10

3. Power Available for Agricultural Field Production — 54

I. INTRODUCTION*

> For a strategy of technological improvement to succeed on a sustained basis, it must include plans to cope with the consequences of its success.
>
> -FAO, 1968 1/

Recent years have witnessed sharply expanded efforts to bring about technological improvement in agriculture in the less developed nations. Primary emphasis has been placed on changes in the technology of production in order to increase food supplies -- in many cases simply to keep output in step with population growth. In view of the historical closeness of the race between "the stork and the plow" in some nations, this emphasis has been well placed. But changes are beginning to take place in agricultural production which will create a need to look beyond physical increases in agricultural output.

A. Need for Broader Perspective

Significant technological breakthroughs have recently been made in the production of key food crops in a few areas of the less developed world. Perhaps the best are the new varieties of wheat and rice which, when combined with appropriate inputs, produce dramatic increases in yields. These varieties have, in turn, stimulated interest in mechanization and other aspects of agricultural development. The success of the new variety package has led some to speak of an agricultural -- or green -- "revolution" in Asia.2/

But revolutions are not tidy affairs; they can have all kinds of dimensions which may not be fully anticipated. Such is the case, at least to some extent, with the new agricultural technologies. Where the breakthroughs have begun to show, a host of new problems are often found waiting in the wings. Some relate to keeping the production advances going; others involve the effects of expanded production on farmers; still others involve problems in income distribution or with by-passed groups. Many are probably not yet apparent. All will need attention at some point.

Technological change in agricultural has traditionally, and perhaps too often, been viewed solely in terms of its immediate impact on increasing food output. Yet technology must be studied in broader terms if second generation problems and issues are to be avoided or mitigated and the pace of agricultural development maintained. While some aspects of technological change in agriculture in the developed world can either be taken

---

* References and notes for this chapter are found on p. 4.

for granted or are well known, the process is considerably less well understood in the many less developed nations of the world -- in part because there have been far fewer technical changes.

What this means is that in planning technological change in agriculture, it is increasingly necessary to move from physical and biological considerations not only to economics, but beyond to social and political relationships. Both quantitative and qualitative aspects need to be considered. Moreover, all these factors must be viewed in terms of their (1) primary and secondary effects, as well as (2) short and long run influence. In total, broad-scale analysis is called for.

B. <u>Interrelations of Technology</u>

Technological change is dynamic. It is both influenced by and influences the environment. On the one hand, as Baranson has noted, feasibility of production techniques is conditioned by the economic, social, and physical aspects of the environment. On the other hand, technology acts as an instrument of environmental change.3/

If we are appraising technical change in terms of economic growth, we similarly find that the relationship is reciprocal -- and that the connections are by no means simple or clear. There are examples of growth causing changes in technology, and changes in technology causing changes in growth. DeGregori suggests that:

> .....not only does growth depend on technological change but also sustained growth is necessary for technological change. Insofar as a new technology requires modification of other elements of the technological complex, continued technological change requires a dynamic economy constantly adjusting to changing circumstances.4/

The relationship, however, does not always lead to an upward spiral: emphasis on economic growth to the exclusion of other factors can lead to serious social costs.5/

Concentration on production technology may also lead to neglect of socio-cultural interrelationships essential to agricultural development over the longer run. These can be of particular importance in less developed nations. It has been suggested, for example, that the lesson of development of Taiwanese agriculture may be that science alone cannot transform agriculture without certain rural organizations being created first or at least concurrently. 6/

As a result of these many interrelationships, agriculture has developed slowly and unevenly. As the Woodruffs have put it: there is no widespread onrush of science and technology in agriculture. Instead, there is a halting movement which is highly localized and connected with special circumstances. Advanced scientific and technological methods in agriculture are, they point out, concentrated in a small part of the world, and often in a small part of a region's or nation's economy.7/

In view of the dynamic nature of technological change, planners and development officials need to consider both the influence of changing environment on the adoption of agricultural technology and the influence of technology on environment.

C. Terms of Reference

The agricultural breakthroughs in Asia have pointed up the need for a broader and more dynamic analysis of technological change. This report will attempt to help meet that need by providing a general review of the nature and effects of technological innovation. Special attention will be given to the impacts of new grain varieties and of mechanization. Implications for public policy will also be examined.

A rather broad interpretation will be taken of agriculture. Both farm and off-farm elements will be included: the supply of inputs, production, and processing and marketing of farm products. While this suggests emphasis on commercial agriculture, the report should be of some relevance to subsistence agriculture.

The report will not, however, focus on the generation of technology. Its availability will largely be assumed. This is not, of course, entirely realistic in the context of less developed nations because the existence of technology cannot be taken for granted. But the generation of technology is an immense subject -- one that is deserving of study in its own right -- and is beyond the scope of this project.

The effects of technology can take many forms. Here we shall speak of two main types: economic, and social/political. Both may make themselves felt at many different levels: farm, local, regional, national, and international. We shall place primary focus on economic effects at the farm and local level. Secondary emphasis will be given to (1) economic implications at the national and international level, and (2) social and political implications at all levels.

We turn first to an examination of the nature of technological change.

References and Notes

1/ "Raising Agricultural Productivity in Developing Countries Through Technological Improvement," The State of Food and Agriculture, 1968, Food and Agriculture Organization, 1968, p. 113.

2/ Lester R. Brown, "The Agricultural Revolution in Asia," Foreign Affairs, July 1968, pp. 688-698; "Green Revolution to Fight Hunger," Washington Post, April 11, 1968.

3/ Jack Baranson, "The Challenge of Underdevelopment," in Technology and Western Civilization (ed. by Melvin Kranzberg and C. W. Pursell, Jr.), Oxford University Press, Vol. II, 1967, p. 517.

4/ Thomas R. DeGregori, review of The Transfer of Technology to Developing Countries, in Technology and Culture, October 1968, p. 632.

5/ For a detailed discussion of this matter in the context of Western civilization, see Ezra J. Mishan, The Costs of Economic Growth, Praeger, Praeger, 1967, 190 pp.

6/ Samuel Pao-San Ho, "Agricultural Transformation Under Colonialism: The Case of Taiwan," The Journal of Economic History, September 1968, p. 327.

7/ William and Helga Woodruff, "Economic Growth: Myth or Reality," Technology and Culture, Fall 1966, p. 474.

II. NATURE OF TECHNOLOGICAL CHANGE*

Technological change may mean different things to different people. No one definition seems to be standard for all disciplines. Here we shall use an economic approach. This definition will be followed by a brief review of the nature of technological change within agriculture.

A. Economic Definitions of Technological Change**

Technology, initially, may be thought of as the broad spectrum of ways or means of carrying out economic activity. It includes "...all those available means which may be used by men to convert scarce natural resources into forms which satisfy human needs."1/

In a more operational sense, economists view such activities in terms of production functions. These functions--which are usually presented in mathematical form--express the role of selected factors in the production of a certain good. Land, labor and capital are the traditional inputs for economic activity. All the technically feasible combinations of these inputs set the limits (or parameters) for the production function.

Technological changes, therefore, may be viewed in terms of (1) modifications of existing production functions or (2) the creation of additional production functions. We will examine both briefly.

1. Changes in Production Functions

Most economists think of technological change in terms of modifications of production functions for existing products.2/ The modification takes place through the addition of new production techniques to the existing stock. New techniques, in turn, may be attained by adding, dropping, or changing at least one of the factors of production. This can mean the adoption of previously unknown or unavailable production or organizational techniques, and the substitution of an existing technique for another. An improvement in quality in an input can also be treated as equivalent to a new factor of production. Many functions are physically possible but relatively few are economically feasible.3/

The optimum economic combination of resources for the production of a commodity will depend on: (1) the form of the relevant production functions, and (2) relative factor and product price ratios. Technological change in agriculture will directly influence the shape of the production function, and can indirectly influence factor and product prices. Technological change in other sectors of society will more directly influence factor and product prices.4/

---

* Reference and notes for this chapter are found on p. 9.

** The general reader may wish to skip this unit and move instead to section B, p. 7.

The development of a modified production function manifests itself in terms of expanding output at the same cost, or producing the same output at lower costs. While some distinction is thence drawn between output-increasing and cost-reducing technological change, this dichotomy may be arbitrary:

> Innovations designed to lower costs ...have a curious propensity to result in expanded output. Indeed it is difficult to conceive of an innovation that successfully lowers costs which does not expand output (and hence lower prices) under free market conditions.5/

In either case, technological improvement must at least momentarily increase the profits (or decrease the losses) of the firm if it is to be adopted. The only exception would be the case where the innovation increased future profit expectations through reduction of risk or uncertainty.6/

Positive technological change may take one of two general conceptual forms. The first is characterized as neutral technical progress: output is expanded without changing factor proportions. The second is characterized as biased technical progress: in expanding output the marginal physical rates of substitution are altered in favor of one factor by specific innovations. In the case of some technologies, both neutral and biased forms may be involved.7/

In any event, complex factor problems may arise. It may not be easy or possible to increase the supply of all factors in the case of neutral change. And the change in factor proportions associated with biased progress may create profound adjustment problems for the disadvantaged factor of production. We shall say more of these difficulties later in this report.

2. Addition of Production Functions

The production function approach is essentially static; among other limitations it is tied to the production of a given item. As one economist recently commented in appraising a study of this type in Peru:

> ...the model assumes no new products, but essentially deals with the problems of producing the old products (including improved versions, however) by better methods.
>
> In some important cases in Africa ...the transformation of traditional agriculture has involved the introduction of new outputs and has then changed incomes rather dramatically.8/

Jones, in fact states that in Africa the greatest technical change of modern times has been the introduction of a complete set of food crops from the Americas.9/

Innovations of this sort can, however, be conceived of in terms of a new production function. In Schumpeter's words:

> ...we will simply define innovation as the setting up of a new production function. This covers the case of a new commodity as well as those of a new form of organization or a merger, or the opening up of new markets.[10]/

This sort of conceptual treatment seems eminently well suited to handling certain forms of technological change.

***

On balance then, we may think of technological advance in agriculture in terms of (1) changes in existing production functions or (2) the establishment of additional production functions. This approach is in tune with one used by Kendrick a few years ago: in defining technical change he included both increases in productivity in resource use, and improvements in the quality and variety of consumer goods.[11]/ In the case of complex technologies, both forms may be involved; and in these and other cases it may be difficult to distinguish between the two.

B. Technological Change Within Agriculture

So far we have discussed technological change in fairly academic terms. How can we more clearly identify the main types of change which actually may be found in agriculture? What are the main characteristics of these changes?

Heady has suggested that technological changes in agriculture can be divided into three main types: biological, mechanical, and biological-mechanical. Biological changes, such as new varieties, have a physiological effect in increasing timeliness of operations, or have some other direct influence on plants or animals.[12]/ Each of the three types might involve changes in existing production functions or the establishment of additional functions.

The various types of technical change can take place at any number of levels of complexity. Bohlen has distinguished three: (1) a simple change in materials and equipment, (2) an improved practice, and (3) an innovation. The simple change involves only variations in accepted behavior patterns. The improved practice means handling two or more variables, but within the general framework of the farmers' general attitudes and activities: e.g. changing from broadcasting to side dressing fertilizer. Innovation requires sharp changes in attitudes and activities: hybrid corn is considered an example because farmers had to realign their values in regard to the source of seed supply and the appearance of the seed.[13]/ While all three types of change are involved in the improvement of agriculture in less developed nations, some of the most striking gains have reflected innovations.

Although technology is generally veiwed in terms of increasing production, it may also take the form of stopping losses. This may be primarily associated with the reduction of losses in storage, but also includes the reduction of production losses (due to insects and diseases) and harvesting losses (due to inadequate labor, etc.). These improvements can involve the same types and levels of change as is true of production.

Exapnded production or reduced losses may lead to quite different factor combinations and levels. Simple biological changes may not, for example, require large quantities of capital; some in fact may even be labor intensive. Mechanical innovations, on the other hand, may require large amounts of capital and lead to labor displacement. Some practices, like double cropping, may require more of both. Both neutral and biased technological change may be involved in agricultural advance.

***

Clearly, technological change in agriculture has many faces and forms. The development of definitions of change that cover this range of activities, yet are both precise and easily understood, is not a simple project. It is an area that deserves further attention.

References and Notes

1/ Nathan Rosenberg, "The Economic Consequences of Technological Change, 1830-1880," in Kranzberg and Pursell, op. cit., Vol. I, 1967, p. 515.

2/ For a dissenting view, see Thomas Balogh, The Economics of Poverty, Macmillan, 1966, p. 75.

3/ Vernon W. Ruttan, "Research on the Economics of Technological Change in American Agriculture," Journal of Farm Economics, November 1960, p. 736; Theodore W. Schultz, Transforming Traditional Agriculture, Yale University Press, 1964, pp. 133, 138; Ruttan, op. cit., p. 736.

4/ F. H. Gruen, "Agriculture and Technical Change," Journal of Farm Economics, November 1961, p. 846.

5/ Ruttan, op. cit., pp. 749-750. Also see Gruen, op. cit., p. 838.

6/ Earl O. Heady, "Basic Economic and Welfare Aspects of Farm Technological Advance," Journal of Farm Economics, May 1949, p. 295.

7/ Ibid., pp. 294, 302; Harry Johnson, "Economic Development and International Trade", in Readings in International Economics (ed. by R. E. Caves and H. G. Johnson), Irwin, 1968, pp. 291-292.

8/ Wolfgang F. Stolper, "Discussion" of "Impact of Technology on Traditional Agriculture: The Peru Case" (by Joseph D. Coffey), Journal of Farm Economics, May 1967, p. 458.

9/ William O. Jones, "Environment, Technical Knowledge, and Economic Development in Tropical Africa," Food Research Institute Studies, 1965 (No. 2), p. 105.

10/ Joseph A. Schumpeter, Business Cycles, McGraw Hill, 1939, Vol. I, pp. 87-88; as cited by Vernon W. Ruttan in "Usher and Schumpeter on Invention, Innovation and Technological Change," Quarterly Journal of Economics, November 1959, p. 598.

11/ John W. Kendrick, "The Gains and Losses from Technological Change," Journal of Farm Economics, December 1964, p. 1065.

12/ Heady, op. cit., pp. 296-297.

13/ Joe M. Bohlen, "Research Needed on Adoption Models," in Diffusion Research Needs, North Central Regional Research Bulletin 186 /January 1968/ (published by University of Missouri), p. 18.

III. ADOPTION PROCESS FOR AGRICULTURAL TECHNOLOGY*

The adoption process for agricultural technology has received considerable study. Most of this work, however, has been done in the developed nations; relatively little seems to have been reported for the less developed nations.1/

A. General Characteristics

The adoption of a new technology is a very uneven process. The rate of acceptance varies geographically -- farmer to farmer, region to region, and country to country -- as well as among crops. But in general the adoption process follows the S-shaped growth curve. (Figure 1)

Figure 1. ADOPTION PROCESS FOR A NEW TECHNOLOGY

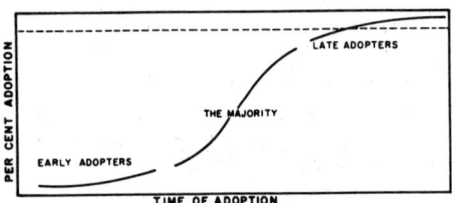

Source: Herbert F. Lionberger, Adoption of New Ideas and Practices, Iowa State University Press, 1960, p. 34.

The combination of variable rate of adoption and the S-shaped adoption curve is vividly represented in the adoption pattern for hybrid corn in the United States. (Figure 2)

Figure 2. PROPORTION OF CORN ACREAGE PLANTED WITH HYBRID SEED

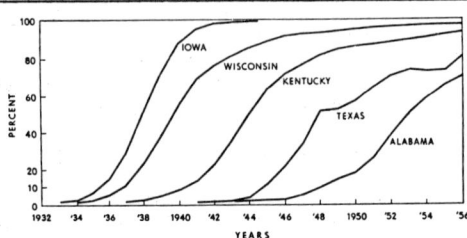

Source: Zvi Griliches, "Hybrid Corn: An Exploration in the Economics of Technological Change," Econometrica, October 1957, p. 502.

---

* References and notes for this chapter are found on pp. 18-20.

Hybrid corn was not a single innovation immediately acceptable elsewhere: the actual breeding of adaptable hybrids had to be done separately for each area. Griliches notes that:

- The lag in the development of adaptable hybrids for particular areas and the lag in the entry of seed producers into these areas can be explained on the basis of varying profitability....

- Where the profits from the innovation were large and clear cut, the change-over was very rapid. In areas where the profitability was lower, the adjustment was also slower. 2/

Whereas the _relative_ superiority of hybrids may have been the same in different areas, the _absolute_ difference in yield in better corn areas was one of the major factors responsible for the differential acceptance rate. 3/

While this analysis was cast in economic terms, Griliches acknowledged that knowledge of sociological variables would be helpful in determining which individual will be first or last to adopt a particular technique. 4/

How do these findings relate to the problems of the less developed nations? We will explore this matter in the remainder of this chapter. Our emphasis will be on the farmer who has left subsistence agriculture, at least in part, and moved into the market economy. 5/

B. Why is Technology Adopted?

New technologies may be adopted for a host of reasons: some obvious, others less so. But in any case, decisions are made at both the farm and governmental level.

1. Farm Level Decisions

The major reasons for farm adoption of new technologies are usually economic in nature, but non-economic variables can sometimes be important.

a. Economic Reasons for Adoption

Direct income improvement would logically seem a prime goal for those using new technologies. Indeed, this is suggested by many studies which indicate that producers in less developed nations respond positively to price incentives. Yet the direct evidence on the motivations for adopting new technologies is slender.

One of the few studies aimed specifically at this point was conducted by Sturt in 1962 in four villages in West Pakistan. His work suggested that cultivators making changes were indeed economically motivated. About 97% of the reasons for making changes were economic. Almost all were directed at obtaining more production; 4% of the reasons were cast in terms of reducing costs and 1% in terms of making things easier. 6/

Why did the producers want more production? The primary reason was to have more food for direct family consumption. The second reason was to have more product to sell or barter in order to make expenditures for consumption or production investments. The consumption expenditures were largely for clothes, followed by food. The production expenditures were about equally divided between bullocks, seeds, and fertilizers.

Unfortunately, little other study of the use of cash income seems to be available. An article by Mohammad suggests that West Pakistan farmers have traditionally had little outlet for savings: the price of land was extremely high and returns on investment were low. Recently, however, many farmers have found tubewells to be a low cost investment with high returns.7/ In India, the AID Mission has reported that while there is little firm evidence as to what is being done with increased incomes, the demand for such inputs as tubewells, fertilizer, high-yielding variety seeds, and machinery (including tractors) is very strong.8/

New technologies also could be adopted for indirect economic reasons -- to take advantage of (or because of) existing resource patterns. The human side could reflect an abundance of, or lack of, management or labor; in some of the less developed nations, for instance, there are seasonal shortages of labor which have spurred the adoption of various forms of mechanization. Or a new technology could be adopted to take advantage of some special resource endowment or previously adopted technology; the adoption of high-yielding varieties of grains, for instance, is facilitated by the presence of irrigation.

In a somewhat different vein, a technology could be adopted by some farmers or in some areas simply to maintain comparative advantage with other areas -- because the others (a) have a greater natural resource endowment, and/or (b) have already adopted a technology which has provided them with an advantage. Growers may not realize the specific reasons, but market forces will help them sense the need to make changes.

b. Other Reasons for Adoption

There are undoubtedly a number of other reasons of a less economic nature which affect adoption of new technologies. We shall focus on only two categories: lack of knowledge, and psychological factors.

Lack of knowledge is a rather negative reason for adoption. Still, many farmers take on new technologies without really knowing whether they make economic sense. This process may represent a calculated gamble, or it may be simply what rural sociologists call "overadoption." Rogers cites three U.S. examples: the use of self-propelled tobacco harvesters in North Carolina, the adoption of hybrid sorghums in Kansas, and the use of four-row corn planters in Indiana.9/ Over the years, a number of technological crazes have swept U.S. agriculture, in part from lack of knowledge.10/

The situation is doubtless less extreme in the less developed nations, if only because low income levels limit adoption of new technology. But at the same time the individual farmer may have less information to form the basis of an enlightened choice. And since many live close to the margin, a mistake could have disastrous consequences. In such a context we might keep in mind Rogers' statement that "...county extension agents may be more effective at preventing the adoption of non-recommended innovations than in promoting the adoption of recommended ideas."11/

Psychological reasons for adopting new technologies are probably more common in developed nations than in developing ones. In developing countries society is more attuned to change: in many less developed nations, tradition is likely to inhibit innovation. In either case, some new technologies may be adopted because the farmer is an innovator at heart. Moreover, the technology may be so new that there is no way to determine its feasibility except by trying it.

2. Governmental Level Decisions

In many societies, the key decisions regarding the implementation of new technologies are made directly or indirectly at the governmental level. The government may influence the availability of inputs (e.g. through import control), the price of products, the nature of technical assistance, etc. As at the farm level, decisions may be motivated by economic and non-economic reasons.

Governments may be largely interested in using new technologies to increase output and provide the basis for increased economic growth. Increased production makes it possible to increase exports and/or reduce imports. It can provide the basis for improving national nutritional levels. The role of agriculture in stimulating economic development, however, has been discussed at length by others and we shall not dwell on it here.12/

Governments can also be influenced by psychological and political factors. Where nationalism is strong, as in many emerging nations, a new technology in agriculture may serve as a symbol of progress. The form can vary, but there must be some dramatic aspect: this means that a more advanced technology may be selected than would otherwise be the case.13/ Tractors have often filled this role, but it is also met by dams, irrigation projects, etc. In some well-known cases -- such as the infamous groundnut scheme in Tanzania (Tanganyika) -- additional drama is provided by the magnitude of failure of the project.14/ But certainly not always.

The political attitudes of the countryside can often have a marked influence on the stability of national governments during periods of political modernization. The attitudes of the peasants are, in turn, strongly influenced by economic conditions. Huntington has observed:

> Someone once said that the glory of the British Navy was that its men never mutinied, or at least hardly ever mutined, except for higher pay. Much the same can be said of peasants.15/

So if a government can improve economic conditions in the countryside, even for relatively short periods, it may reduce political instability. To the extent that technological innovations can make an economic contribution to the countryside, then, they may be favored by the government. But as we shall see in later chapters, this can be a treacherous course and may not work out quite as anticipated.

C. Factors Influencing Rate of Adoption of Technology

Perhaps the major factor influencing the rate of adoption, as the previous portions of this chapter have suggested, is the real or anticipated profitability of a technology.[16] In this section we move on to other characteristics influencing adoption. These include (1) the technology itself, (2) the adopter, (3) the economy, and (4) the society.

1. Characteristics of the Technology

Technologies may represent varying degrees of complexity -- from a slight modification of current practices to a broader scale innovation. Obviously those which are least complex are apt to be adopted first (unless the payoff is so minor that no one bothers with it). Similarly, a technology which is divisible and can be adopted in parts or stages is apt to find more rapid use.

A study of the adoption of four farming practices in Japan indicated the following attitudes and rate of use:[17]

Practice	Favorable Attitude	Tried	Adopted
	------ percent ------		
1. Testing soil for fertilizer needs	96	82	73
2. Seeding recommended amounts of rice	84	78	66
3. Using Norin No. 30 Soybean seed	57	38	33
4. Drying soil and planting two crops a year	67	21	12

The first two practices called only for slight changes in techniques; they were favorably received and widely tried and adopted. The use of a new soybean seed was less favorably received, but most of whom tried the practice adopted it. Drying of soil and planting of two crops was more favorably viewed than the seed, but a much smaller portion both tried and adopted the practice. The latter practice would have called for basic changes in farm operation.

Similarly, a study of the adoption of soybeans in the Yaqui Valley of Mexico revealed that the compatability of soybean production with existing practices, and the ease with which it could be carried out, were decisive factors in its rapid adoption by the farmers of the region.[18]

Other factors influencing rate of adoption include communicability and technical "appropriateness". The latter, as defined by Byrnes, is the relevance of the innovation or practice for the particular farmer, given his educational, social, agricultural, and economic situation.[19]

2. Characteristics of the Adopter

Those who are the most rapid to adopt new technology generally have similar characteristics. These might be classed as economic (or situational) and social (or personal).

As might be expected, the farmers who are most apt to adopt new practices are among the more economically favored: they are relatively well-to-do and have large operations. They can afford to take the risk involved in adopting the technology--and at the same time stand to reap the most economic benefits.[20] (The related question of land tenure, however, is a more cumbersome one: there is no categorical answer except that owner-operators are generally more willing to invest in their farms than absentee landlords.)[21]

The nature of the farmer's cropping pattern is also of economic influence. If the good produced is of relatively high income elasticity of demand and designed for export, it may be grown under more technologically advanced conditions than if it is destined for local markets. Mexico provides a well documented case: the greatest growth has been achieved on (1) irrigated export crops -- especially fruits and vegetables -- produced on private lands in the Pacific North Region, and (2) relatively extensive crops such as cotton and wheat grown on a small number of large scale farms. Change has proved most difficult in Mexico where a traditional domestic crop such as corn is grown intensively near the margin.[22]

Social characteristics of adopters may be closely linked to their economic status. Those who make technical changes are more apt to be well educated, socially active, and have more contacts outside the village than is true of those who do not. As Bose indicated in 1961, following a study in Indian villages: "...those who adopted more belonged to higher castes, were literate, and had higher participation in community activities." Similar findings have been obtained in a more recent study in India.[23]

Sociologists place considerable emphasis on the role of the interaction effect in spreading technology. It is "...the process through which individuals in a social system who have adopted an innovation influence those who have not yet." In certain Oriental or tribal societies, social and cultural conditions of rural life may also have a significant influence on the introduction of new techniques.[24]

3. Characteristics of the Economy

The rate of adoption of technology will also be influenced by the nature of the economy: the nature of the infrastructure, the demand for agricultural products, off-farm employment, and government policies. [25]

Infrastructure includes the availability of inputs necessary for the change, the availability of credit, and the nature of the marketing system (communication, transportation). It might also be said to include research and education. In most less developed nations, these vital ingredients are all seriously lacking to various degrees.

Demand for the products produced or marketed under the influence of new technologies will play a key role in determining their profitability, and hence as their rate of adoption. High grain prices in India during the mid 1960's, in part due to drought, undoubtedly accelerated the adoption of improved cultural practices. Demand, however, needs to be considered in terms of both domestic and foreign markets.

Off-farm employment can play an important role in tempering the pace of technological improvement. Labor-saving technologies, for instance, will be of less merit if there is no place for displaced labor. In Japan, technological development of agriculture has been accelerated because of the availability of off-farm employment: the opportunity cost of labor was raised, providing greater incentive to adopt new practices.

Government policies can have a most significant impact on the adoption of technologies. The rapid increase in private tubewells in West Pakistan during the 1960's, for instance, was aided by (1) higher and more stable prices for agricultural products, (2) lower cost power as a result of the government's electrification program, and (3) increased availability of pump materials due to the import liberalization program. Pakistan, as do other nations, also provides subsidies on the cost of inputs such as fertilizer, crop protection, and improved seed.26/

D. **Possible Rejection of Technology**

Few new technologies, even when adopted or used once, are permanent. They may eventually be replaced by more improved technologies or they may gradually or suddenly be dropped. There are many factors influencing the partial or total rejection of a technology. Here we shall note only a few.

There are numerous instances of technologies being introduced without adequate research on some key phase of production or marketing. Miracle cites three examples in Africa:27/

> --rejection of chemical fertilizers by farmers in Eastern Nigeria because the fertilizers introduced caused their rice to develop too much straw and their yams to store poorly.
>
> --the abortive attempt to introduce pigeon peas and soybeans to the northern Congo Basin without research into how pulses are processed by people involved.
>
> --attempts to introduce improved maize varieties in the Congo Basin that failed because of lack of research on consumer preferences as to hardness and color of the grain.

A related problem is that production research may have been conducted under conditions which cannot be equaled by the average farmer. Demonstration plots are a way around this problem.

Misapplication is probably a major reason for rejection of technologies. Once again there are probably endless examples, but one from a Community Development Program in India suggests the major difficulties:

> In this particular village the government workers had already urged the farmers to try some fertilizer. They applied it too liberally, and the crops withered and died. Next year, the same villagers, still friendly, accepted the advice to plant wheat in an empty irrigation reservoir. Rust attacked the crop. After that the men ruined an expensive German sprayer in an effort to kill the rust. Government officials ended up by regarding the peasants as hopelessly stupid and lazy. Peasants who could not afford to risk their crops stuck to traditional ways they knew would work after a fashion.[28]/

Misapplication, it will be noted, can come about because of failures or inadequacies of all parties involved -- from government official down to peasant.

Farmers may try new techniques -- such as high-yielding varieties of grain -- then turn from them because of shortages of inputs such as water, fertilizer, or insect and disease controls. Reding, for instance, found that Mexican growers who dropped hybrid corn were generally those without irrigation.[29]/

Moreover, a technology which may seem profitable under one set of economic conditions may become uneconomic under another. An increase in costs and/or a decrease in returns can lead to discontinuance, or more limited use, of a technology. Price declines brought about by production increases may limit grower interest in new varieties of rice. Government taxation policies can also play a role: Chauhan has reported the case of a village in India which had -- after learning new methods of cultivation -- adopted the cultivation of a new crop (tobacco), but which reverted to traditional crops (cotton and maize) following the third increase in taxes.[30]/

All or part of the preceding factors may interact to discourage use of technology. So might the lack of complementary inputs. These points may not be specifically recognized by the farmer, who only knows that the technology isn't working out the way it was expected. But the result is the same. Many of the problems may be worked out over time, but changing economic relations remain a constant threat. The degree to which a farmer retrenches will be influenced by his technical and capital resources, and his outlook.

***

These, then, are some of the major characteristics of the adoption -- or rejection -- of agricultural technology.

References and Notes

1/ AID has, however, sponsored studies under the direction of Everett Rogers of Michigan State University in India, Brazil and Nigeria. The India studies have been published and will be noted in this chapter.

2/ Reproduced from Zvi Griliches, "Hybrid Corn: An Exploration in the Economics of Technological Change," Econometrica, October 1957, p. 522.

3/ Zvi Griliches, "Congruence Versus Profitability: A False Dichotomy," Rural Sociology, September 1960, p. 354. Some other comments pertaining to the relative roles of economic vs. sociological variables unleashed an extended debate in Rural Sociology. See the following issues: December 1959, pp. 381-383; September 1960, pp. 354-356; December 1961, pp. 409-414, September 1962, pp. 327-332.

4/ Griliches, op. cit. (1957), p. 522.

5/ Technological change in subsistence economies is a somewhat special matter involving more attention to risk and survival than is true of the cases discussed here. For a perceptive and comprehensive review of this matter, see Clifton R. Wharton, Jr., "Risk, Uncertainty and the Subsistence Farmer: Technological Innovation and Resistance to Change in the Context of Survival," Agricultural Development Council, December 1968, 60 pp.

6/ Daniel W. Sturt, "Producer Response to Technological Change in West Pakistan," Journal of Farm Economics, August 1965, pp. 630, 632-633. A more general survey of growers wanting to make changes gave the following breakdown: increase production, 71%; less effort, 21%; reduce expenses, 7%. (p. 632)

7/ Ghulam Mohammad, "Private Tubewell Development and Cropping Patterns in West Pakistan," The Pakistan Development Review, Spring 1965, p. 45. Also see W. P. Falcon and Carl H. Gotsch, "Lessons in Agricultural Development - Pakistan," in Development Policy - Theory and Practice (ed. by G. F. Papanek), Harvard University Press, 1968, pp. 272-278.

8/ "India Program Memorandum, FY 1970," US/AID Mission, New Delhi, Annex D, September 1968, p. D-66.

9/ Everett M. Rogers, Diffusion of Innovations, The Free Press, 1962, p. 144. For details on the tobacco harvester case see: W. D. Toussaint and P.S. Stone, "Evaluating a Farm Machine Prior to its Introduction," Journal of Farm Economics, May 1960, pp. 241-251.

10/ For examples, see: Arthur H. Cole, "Agricultural Crazes," American Economic Review, December 1926, pp. 622-639; Earl W. Hayter, The Troubled Farmer, 1850-1900; Rural Adjustment to Industrialism, Northern Illinois University Press, 1968 (see index).

11/ Rogers, op. cit., p. 145.

12/ See, for example, John W. Mellor, The Economics of Agricultural Development, Cornell University Press, 1966, pp. 3-130 (Part I).

13/ Baranson, op. cit. (see fn. 3, chp. I), p. 520.

14/ See Alan Wood, The Groundnut Affair, The Bodley Head (London), 1950, 264 pp.

15/ Samuel P. Huntington, Political Order in Changing Societies, Yale University Press, 1968, p. 374. Also pp. 291-300, 374-376.

16/ A sociologist suggests that "what really determines the rate of adoption of an innovation is the adopter's perception of profitability and not objective profitability" (Rogers, op. cit., 1962, p. 140).

17/ David E. Lindstrom, "Diffusion of Agricultural and Home Economics Practices in a Japanese Rural Community," Rural Sociology, June 1958, pp. 174-175. A somewhat similar breakdown is provided for India in Prodipto Roy, et al., Agricultural Innovations in Indian Villages, National Institute of Community Development (Hyderabad), 1968, pp. 13-24.

18/ Abdo Magdum M. "The Diffusion and Adoption of Soybean Cultivation in the Yaqui Valley," in Communications in Agricultural Development (First Inter-American Research Symposium, October 1964), ed. by D. T. Myren, Mexico City, p. 142.

19/ Francis C. Byrnes, "Some Missing Variables in Diffusion Research and Innovation Strategy," Agricultural Development Council (New York), ADC Reprint, March 1968, p. 2.

20/ Tagumpay-Castillo, however, suggests that smaller farmers have a greater propensity to adopt new practices during "the trying out" period. But their unwillingness and inability to borrow money for the costs incident to full adoption limit the extent of adoption. (Gelia Tagumpay-Castillo, "Propensity to Invest in Agriculture. Observations from a Developing Country - the Philippines," International Journal of Agrarian Affairs, July 1968, p. 309.)

21/ Santi Priya Bose, "Characteristics of Farmers Who Adopt Agricultural Practices in Indian Villages," Rural Sociology, June 1961, p. 138; Lindstrom, op. cit., p. 181; Lionberger, op. cit., p. 34; Rogers, op. cit., p. 175 ff.; Roy, op. cit., pp. 30-44, 100; Tagumpay-Castillo, op. cit., p. 309.

22/ Reed Hertford, "The Development of Mexican Agriculture: A Skeleton Specification," Journal of Farm Economics, December 1967, p. 1175; Bruce F. Johnston, "Agriculture and Economic Development: The Relevance of the Japanese Experience," Food Research Institute Studies, 1966 (No. 3), p. 286; W. Whitney Hicks, "Agricultural Development in Northern Mexico, 1940-1960," Land Economics, November 1967, pp. 401-402.

23/ Lionberger, op. cit., p. 34; Bose, op. cit., p. 138; Roy, op. cit., pp. 45-54, 101.

24/ Rogers, op. cit., p. 138. Also see George M. Foster, Traditional Cultures and the Impact of Technological Change, Harper, 1962, 292 pp.

25/ For a recent study in variation in rates of adoption between villages in India, see Frederick C. Fliegel, et al., Agricultural Innovations in Indian Villages, National Institute of Community Development (Hyderabad), 1968.

26/ Falcon and Gotsch, op. cit., p. 277. For information on fertilizer subsidies in various nations, see Fertilizers: An Annual Review, 1967, Food and Agriculture Organization, 1968, pp. 36-44.

27/ Marvin P. Miracle, Agriculture in the Congo Basin, University of Wisconsin Press, 1967, p. 288.

28/ Barrington Moore, Social Origins of Dictatorship and Democracy: Lord and Peasant in the Making of the Modern World, Beacon Press, 1966, p. 401. For background on the program see Douglass Engminger "Overcoming the Obstacles to Farm Economic Development in Less Developed Countries," Journal of Farm Economics, December 1962, pp. 1376-1377.

29/ Jesus M. Reding, "Social and Economic Factors Which Influence the Diffusion and Adoption of Hybrid Corn in the Bajio," in Communications in Agricultural Development, op. cit. (see fn. 18, this chapter), pp. 132-136.

30/ Brij Raj Chauhan, "Rise and Decline of a Cash Crop in an Indian Village," Journal of Farm Economics, August 1960, pp. 663-666.

IV. IMPACT OF CHANGES IN AGRICULTURAL TECHNOLOGY*

The effects of new technologies are normally first felt at the farm level, and then spread through the rest of society. We shall follow the same pattern here -- starting with a discussion of the impact of technologies at the farm and then turning to effects at the national and international levels.

A. Impact at the Farm Level

The effects of technological advance may make themselves felt in many ways. We shall look at two categories: economic and social/political. Primary emphasis will be placed on the former.

1. Economic Implications

Technologies have both primary and secondary economic effects. The primary effects are more obvious than the secondary effects, but not always more important in the long run.

a. Primary Effects of Technology

The most obvious direct effects of agricultural technologies are in terms of supply and farm income, but they may also provide a powerful influence on employment and the use of other technologies.

(1) <u>Influence on Supply</u>. Most technological changes in agriculture have the effect of increasing the supply of agricultural products. Increased supply is, of course, largely traceable to improvements in production practices, but may also be related to improvements in storage. Both factors, especially the latter, may also have an effect on the quality of the product.

Examples of production increases are legion. Perhaps the most striking recent example has been, as we have suggested, the adoption of new high-yielding varieties of grains. Related multiple cropping practices, often associated with mechanization, offer further possibilities of sharply increasing output per unit of land. These innovations will be discussed in considerable detail in the next two chapters.

The need for improved storage and marketing systems in reducing losses in less developed countries has been widely accepted. Kriesberg has cited the following examples of losses:

> -In India, insects reportedly cause post-harvest losses of at least 10% of cereals; rodents are reputed to cause an added loss of 10 to 20% of stored grains.

> -In Syria and Lebanon, losses of 10 to 20% have been reported for stored grain.

---

* References and notes for this chapter are found on pp. 32-34.

-In Brazil, a survey of processors, merchandisers and
producers placed losses of stored grain at 15-20%.

By comparison, he reports that the massive grain storage program in the
U.S. operates with losses of only 0.5% annually.1/ Much attention is
being given to these matters, but there is a wide range for improvement --
especially for more perishable commodities.

In addition to increasing supply, improved technologies can also reduce
variability in supply. This is obvious for storage but may also be
true in terms of production. Shaw and Durost, after studying changes in
the U. S. corn belt, reported that the use of better varieties of corn
and improved cultivation and fertilization practices reduced variations
in yields in good and bad weather.2/ It is not certain how much has been
accomplished along this line in the less developed nations, but irrigation
and multiple cropping have undoubtedly had some effect.

(2) Influence on Farm Income. If the effect of many new
technologies on increasing production is clear, their influence on farm
income is considerably less so. While our discussion in this section will
be necessarily based on the situation in the developed world, the basic
points should have relevance for the developing nations.

The income gains to agriculture are relative in nature: they come to those
who first adopt technological advances but are competed away over the
longer run. As Wilcox and Cochrane put it, the operators who first adopt
a new technology reap the income benefits (the difference between the old
price and new lower costs). The first farmers undertake a new method or
practice for the obvious reason that they benefit directly.3/

Within the innovative group, it is possible that those who quickly follow
the pioneers -- the early imitative firms, as Kendrick calls them --
could realize even greater returns because of the fewer problems involved
in the application of more perfected innovations. Wilcox and Cochrane go
on to note that the higher profits of the superior farm managers tend to
result from their ability to single out the profitable new techniques and
to adopt them promptly.4/

The amount of profit received by the early adopters depends in part on the
rate of adoption of the innovation: the faster the diffusion of an
innovation, the smaller the total abnormal profit. In the farm sector,
competitive conditions are such that the time lag and the rate of profit
accruing to the farm sector for innovations is kept to the absolute
minimum. The latter point, however, may be more true of developed than
less developed economies.5/

As more and more farmers adopt the new technological advance, the income
situation begins to change character. Two things happen: (1) in the
usual case where output is significantly increased, the price of the
commodity falls;6/ (2) in the less common case where costs are reduced
(output remaining the same), the gains from the new practice or technique
are capitalized into the value of the fixed asset involved. Consequently,

in the long run, by the time most farmers have adopted the technology, the income benefits realized by the first farmers have disappeared.7/

If this is the case, why do later groups of farmers follow in adopting the innovation? The reason, as Wilcox and Cochrane have expressed it, is that they are caught on a technological treadmill. As the technology is more widely adopted, and production increases, prices go down; yet costs for non-innovative farmers do not decrease. Thus the farmer who does not adopt new technology is squeezed. To stay even with the more progressive farmers he is forced to adopt the new technology. The position of the farmer who is not able to make a change -- because of lack of capitalization or inapplicability to specific farming operation -- is not a fortunate one. He is clearly disadvantaged.8/

The treadmill concept, however, may not be fully appropriate for less developed societies. This is because, in part, the non-adopters are more likely to be self-sufficient and not so affected by market pressures. The worst that may happen to them in an absolute economic sense is that they are bypassed by progress: this has, for instance, clearly been the case of the small farms or "ejidos" in Mexico.9/ Also, increased production could merely replace imports with minimal adverse effect on price. Still, the concept provides a useful conceptual starting point.

The type of influence that technical change has on the net revenue of agriculture will depend to some extent on the type of innovation and the price elasticity of demand. In an earlier chapter we referred to Heady's categorization of innovations into three main types: biological, mechanical, and biological-mechanical. In the same reference he related these to elasticity of demand, made allowance for differing effects on output and cost, and then offered conclusions on the effect on net revenue of agriculture. His analysis is summarized in Table 1.

While the various combinations can result in differing effects on income, some are more probable than others. One of the most likely combinations, for instance, is a biological-mechanical innovation influencing a crop with inelastic demand where output and costs increase: this results in a net decrease of revenue. But the demand for the products of an individual farm is more elastic than for the industry, so that the effect at the farm level may be more favorable than for agriculture as a whole.

On balance, we can see that the adoption of technical advances makes some farmers better off, and some worse off. An important economic question, though, is whether the farmers in the latter category are worse off in a relative or in an absolute sense. If their income position remains the same while that of others improves, they are **relatively** disadvantaged; if their income decreases, they are also **absolutely** disadvantaged. The number of farmers who are relatively disadvantaged by a technological innovation undoubtedly exceeds those who are absolutely disadvantaged.

We should not, however, limit our analysis only to existing practices. As we have noted earlier, new technologies may make it possible to introduce new income-earning products and services.

Table 1. EFFECT OF TECHNOLOGICAL CHANGE IN AGRICULTURE ON NET FARM INCOME UNDER VARYING CONDITIONS

Type of Innovation	Price Elasticity of Demand	Effect of Technological Advance on Total		Effect on Net Revenue of Agriculture
		Output	Cost	
Biological	Elastic	Increase	Increase	Increase or decrease 1/
	Inelastic	Increase	Increase	Decrease
Mechanical	Elastic	Constant	Decrease	Increase
	Inelastic	Constant	Decrease	Increase
Biological - Mechanical	Elastic	Increase	Increase or Decrease	Increase or decrease 1/
				Increase
	Inelastic	Increase	Increase or Decrease	Decrease
				Increase or decrease 2/

Notes: 1/ Total revenue will be more. Net revenue will increase if the increase in total revenue is greater than the increase in total costs. Net revenue will decrease if total revenue is increased by a smaller amount than the increase in total costs.

2/ Total revenue will be less. Net revenue will increase if the decrease in total revenue is less than the decrease in total costs. Net revenue will decrease if the decrease in total revenue is greater than the decrease in total cost.

Source: Adapted from Earl O. Heady, "Basic Economic and Welfare Aspects of Farm Technological Advance," Journal of Farm Economics, May 1949, pp. 298-299.

For example, the tropical latitudes, high elevation, and sufficient rainfall encountered in certain regions of Kenya were found to be particularly suited to growing the pyrethrum flower. Pyrethrum provided a cash crop to help offset the risk associated with the cultivation of coffee and tea; it did not require the years of investment prior to productive return nor did it involve the hazard of world price fluctuations. Continued research to expand the end use of pyrethrum have reduced processing costs and have helped to create a world market (over $10 million worth sold as insecticide) that has expanded threefold during the period 1958-1961.10/

In this sort of situation, the outcome is likely to largely be both absolute and relative advantage, with only limited relative and absolute disadvantages.

In any case, the effects of technological change go beyond their immediate effect on income at the farm level.

   (3) Influence on Employment. New agricultural technologies can have both positive and negative impacts on labor. Mechanical innovations in agriculture as well as industry, it is widely recognized, generally result in displacement of labor. In fact, Kendrick suggests that the major absolute loss may be the earnings lost by those unemployed persons who lost their jobs because of technological change.11/

For those employees who remain in agriculture, incomes are not necessarily reduced. It depends on the impact of the technology on output (and in turn on prices and revenue) and costs. Wilcox and Cochrane point out that the average labor income of those remaining in agriculture derives from a three-sided struggle between declining total revenue, declining total costs, and declining number of workers. The displacement of labor is a particular problem with nations which are lower on the ladder of development and have limited off-farm employment opportunities.12/

But there are examples of agricultural development stabilizing or increasing the demand for labor. Technical improvements in Japanese agriculture, for example, led to increased double cropping: peak work loads for the new crops were scheduled to coincide with slack periods for the old, resulting in a more even spread of work during the year. The same was true in Taiwan. In other areas new technologies are setting the stage for new and more intensive types of agriculture. As Mellor suggests, there is considerable evidence in most low income countries that technological advance requires a complementary input of labor.13/

  b. Secondary Effects of Technology

The secondary economic effects of technological advances at the farm level are often no less important that the primary effects.

Economists have, for instance, tended to view increases in production in terms of short-run demand interrelationships. Longer-run supply reactions, as Gruen has noted, have been relatively overlooked. Technological change can have expansion and substitution effects. The <u>expansion</u> effect is the encouragement to switch resources from other avenues of production. The <u>substitution</u> effect sets in after production has increased and prices are reduced, and influences both demand and supply.14/

As an example of these effects, Gruen cites the example of virus disease which was introduced into Australia in 1950 and killed about 95% of the rabbit population. This occurred at a time when they provided a major obstacle to increases in wool production. Reduction of rabbit numbers made it possible and profitable to expand wool production: the expansion effect was to redirect available investment funds to wool production. As a consequence, wool prices declined, setting in motion the substitution effect: on the demand side, Australian wool consumption was encouraged, possibly at the expense of the rate of growth of rival wool suppliers and synthetic fibers; on the supply side, Australian farmers started switching from wool to other products, some of them outside agriculture.15/

The influence of technological change on supply and demand relationships may materially change existing patterns of comparative advantage -- between crops, farms and areas.. Severe disruptions can result, as is illustrated by the example of the Colombian cotton industry. The introduction of American upland cotton, grown under modern conditions, proved too much for the native types of cotton grown under traditional conditions: the old "colonial" sector of the cotton industry was progressively eliminated and in its place, a new type of production unit was built, with its own special technological and organizational weaknesses.16/

Technical changes can also lead to pronounced changes in farming organization. Perhaps the most pronounced change is the impetus it gives to moving from a non-market to a market orientation.

> Adoption of new technology has usually increased agriculture's dependence on the non-farm sector of the economy and has lessened agriculture's dependence upon land inputs.17/

In order to obtain these inputs -- such as machinery, fertilizer, etc. -- the farmer needs cash or its equivalent. This can generally only be secured by selling or exchanging products.18/

A slightly different impact may take place with respect to structure. Depending on the existing type of agriculture, technological improvement may result in (1) a shift from extensive to intensive production, or (2) an expansion in farm size. Herr suggests that in Australia, technological changes resulted in a shift from extensive grazing to intensive livestock and agricultural production; in the United States the principal reaction was an increase in size.19/

Furthermore, technological change in one sphere of agriculture can influence the need for -- or opportunity for -- technological change within other spheres. Not only are improved production techniques needed, but marketing will need to be improved. Thus the introduction of one technology may set into motion a whole train of related technologies.

2. Social/Political Implications

Technological changes at the farm level may well have significant social and political implications. This is a matter that can, and probably should, better be explored by others. Still, a few observations may not be out of order.

The social impacts may take several forms. Perhaps the best known is the influence on rural social structure. As Forbes has put it:

> Changing the type of agriculture literally changes the way of life, for agriculture is deeply rooted in the material civilization of these /the underdeveloped/ countries and rapid, basic change can offset its undoubted benefits with a profound disruption of social fabric.[20]/

Another social problem stems from the fact that some farmers are advantaged while other farmers may be disadvantaged -- both relatively and absolutely. The situation of the rural unemployed may be particularly unfortunate. In any case, the result can be a growing gap between certain sectors of the agricultural community.

Such impacts are not likely to guarantee economic or social stability in the countryside. Moreover, the rural under-employed may be forced to move -- as in the United States -- to the cities, exacerbating already serious urban problems. We shall return to the national aspects of these problems later in this chapter.

B. Impact at the National Level

The effects of technical change in agriculture extend well beyond the farm. In fact, the most lasting benefits fall to consumers as a whole. As at the farm level, these effects may be primarily economic or social/ political in nature. But biological and/or ecological implications should also be noted.

1. Economic Effects

The main economic gain from new technology ultimately falls to consumers in the form of increased output of agricultural goods at lower prices. In addition to increases in quantity, quality (including nutritional value) may also be improved. These contributions form the basis for economic growth.

As we have noted, at the early stages in the introduction of an innovation, the impact on quantity is not great. It is only as the technology is more widely adopted that there is a significant increase in supply. Thus it is the masses who follow the first few farmers who:

>...are the ones who make the greatest absolute contribution to lessening the real price of food and to freeing resources from agriculture. Yet these producers are promised negative payoffs or costs for the contribution, because their incomes are reduced from the process under inelastic demand.21/

Just how closely this process will be followed in the less developed nations remains to be seen, but it does provide a useful starting point.

In any case, it may seem that there is an intra-industry transfer of income. (Measure needs to be taken of the public resources devoted to agricultural research and education, irrigation, roads, subsidies on inputs such as fertilizer, etc.) This transfer may lead to conflicts in public policy because, as Heady points out, groups which sacrifice from aggregate change are likely to have different policy preferences than those who gain from the overall change.22/

The gradual -- and sometimes sudden -- shifts in comparative advantage that result from technical change can set the stage for serious adjustment problems of broad concern. We have seen this in the United States where improvements in transportation provided low freight rates for the agricultural products of the American Midwest and led to the decline of New England farming.23/

More recently, as a result of agricultural modernization in India, comparative advantage is shifting to the northern portions (the Ganges plains) and to the major rice bowls of the southern parts. As a result, Schultz points out:

> A very large triangle in Central India is losing out competitively. Scores of millions of people who are dependent upon agriculture reside in this area that will be left behind.24/

And within these regions in India there are gaps between the large, relatively modern farms and the small family holdings. Moisy states that there is a risk of famine pockets in the regions bypassed in the development process, partly because of an inadequate marketing and distribution system.25/

2. Social/Political Effects

Technological change can have social and political ramifications at the national level. They may range from moderate to severe in effect; the determination depends in part on who is affected and how.

India offers a clear example. As a result of recent technological changes, the middle class in agriculture in some regions has been strengthened while the landless laboring class has not found its position significantly improved. The middle class group is reacting by demanding:

- a continued flow of yield-increasing technology
  and related inputs

- adequate price-cost benefits

- better schools, better housing, improved rural
  roads and transportation, electricity etc.

They are backing up their demands by increased participation in village, district, state, and national political affairs.[26]/ The less advantaged group is also increasing their participation in political matters, but often in a less orderly way: through uprisings and unrest.

Peasant reaction to modernization will, as Moore indicates, be strongly influenced by the types of social organization, and the timing and character of the modernization process. Their response may be passive or it can go as far as revolutionary potential.

> Whether or not this potential becomes politically
> effective depends on the possibility of a fusion
> between peasant grievances and those of other strata.
> By themselves the peasants have never been able to
> accomplish a revolution.[27]/

The social and political ramifications of most technological changes in agriculture are, of course, of a much more modest level. But the point is that some effects of this nature can be expected from almost every innovation -- and in some cases these can be of significance at the national level.

3. Ecological Effects

One aspect of technical change that has been almost completely overlooked by most contemporary observers is its impact on ecological balance. Such an effect, however, is more likely seen in retrospect.

Jones has noted the ecological influence of the cultivation of maize in Africa:

> Although this crop will outperform the millets and
> sorghums when rainfall is sufficient and properly
> distributed, it suffers more from deficiencies in
> moisture supply; farming communities that have
> shifted heavily to maize may find that the incidence
> of famines, or its severity, has increased. Culti-
> vation of maize in pure stand can also result in
> heavy losses of soil through erosion.[28]/

Similarly, Ripley has observed that expanding human populations and improved veterinary medicine are causing widespread overstocking of range lands with numbers of livestock far in excess of the capacity of the natural vegetation to support them. He adds that the downward trend in productivity of the world's arid lands is indisputable, and that it is accelerating at an unprecedented rate.29/

But technology can do as much or more to improve conditions as its unwise use can lead to deterioration. It is a two-edged sword and those responsible for its use should keep this in mind.30/

C. Impact at the International Level

The effects of technological change can show themselves at the international level in both obvious and less obvious ways. We shall note two of the most important.

1. Comparative Advantage and New Products

Technical change can lead to sharp changes in international comparative advantage for agricultural economies -- both in terms of production of traditional products and the introduction of new products.

History provides a number of examples. The development of the steamship and railroads, for example, opened up vast areas in the United States and Oceania and made it possible for them to provide food and raw material products to Great Britain at prices with which British agriculture could not compete: this led to a sharp decline in the agricultural sector in Britain.31/ Innovations have also, as Baranson has pointed out, provided the basis for agricultural commodities entering world trade:

> ...cases in point are a decortication process to reduce the bulk of sisal hemp, a milling process to prevent deterioration of sugar cane, the sulfate process for wood pulp, and the mechanical cream separator that commercialized Danish butter.32/

Similar innovations might work to the advantage of present or potential products of the less developed nations.

Other examples are less dramatic. But it should be recognized, however, that developed countries -- as well as less-developed -- may have comparative advantage tilted in their favor by technological change. In other words, the less developed nations may have to step up use of technology just to stay even in the international market place.

2. Role of Export Industries

Foreign investments in plantations and other export-oriented industries have been an important source of technological change.

The degree to which such innovations have had an impact on the national economy is a moot point. It depends on the degree to which the foreign firm is integrated into the economy. If it is a foreign-owned and managed plantation producing exclusively for export -- in essence an enclave -- the influence may be minimal. But if it influences production on farmer-owned units which is purchased under contract, the influence on the national economy may be somewhat more substantial.

Either case, however, presents further analytical difficulties. The problem lies in the potential export of "consumers surplus." The economic advantages of new agricultural technologies primarily accrue to consumers in the country where they are produced. Thus the technologies work to the national -- if not to the farmers' -- interest in the long run.

The situation is changed, however, if the products influenced by the technology are exported. The consumers are not at home but abroad. The "consumers surplus" is not retained but is lost to the producing nation to the extent that the product is exported. Thus Singer indicates that:

> ...The main requirement of underdeveloped countries would seem to be to provide for some method of income absorption to ensure that the results of technical progress are retained ...33/

Furthermore, this absorption needs to be re-invested. This means that

> ...a flow of international investment into the underdeveloped countries will contribute to their economic development only if it is absorbed into their economic system; i.e., if a good deal of complementary domestic investment is generated and the requisite domestic resources are found.34/

These considerations begin to take us into complex matters of trade and development policy and are beyond the scope of this report.35/ But as foreign investment can be an important source of technology, the matter needs to be considered at the policy level.

D. Interrelationships

The separation of the effects of technological change by farm, national, and international levels has, of course, been arbitrary. The three are highly interrelated. A change at the farm level, like a stone tossed in water, sends out waves in all directions. The economic stages, for example, have been summarized as follows:36/

(1) Effect on the cost structure or product mix of the individual farm in which new techniques are adopted.

(2) Shifts in industry demand curves for factors of production and supply curves of final products.

(3) Change in rate of growth and distribution of total and per capita income or leisure in the whole economy.

The challenge for policy makers is to determine the various economic, social, and political interrelationships and to take steps to capitalize on the benefits and minimize the possible disruptions.

***

In the next two chapters we shall review the interrelationships of technology in terms of two relatively recent and important innovations in the less developed world: new high-yielding varieties and mechanization. The discussion of each will follow the chapter sequence used so far in this report: introduction (background), nature of technological change, adoption process, and impact of changes. Hopefully this process will provide not only empirical example but also a framework which will be of value in analyzing other technological changes.

References and Notes

1/ Martin Kriesberg, "Marketing Food in Developing Nations -- Second Phase of War on Hunger," Journal of Marketing, October 1968, p. 58.

2/ Lawrence H. Shaw and Donald D. Durost, The Effect of Weather and Technology on Corn Yields in the Corn Belt, 1929-62, U. S. Department of Agriculture, Agricultural Economics Report No. 80, July 1965, p. iv.

3/ W. W. Wilcox and W. W. Cochrane, Economics of American Agriculture, Prentice Hall, 1960, pp. 325-326. Also see Earl O. Heady, Agricultural Policy Under Economic Development, Iowa State University Press, 1962, p. 10.

4/ Kendrick, op. cit. (see fn. 11, chp. II), p. 1068; Wilcox and Cochrane, op. cit., p. 52.

5/ Kendrick, op. cit., p. 1069; Wyn F. Owen, "The Double Developmental Squeeze on Agriculture," American Economic Review, March 1966, p. 55.

6/ And in the general case where the price elasticity of demand at the farm level is less than 1.0 (that is, demand is inelastic), gross returns to all producers must fall.

7/ Wilcox and Cochrane, op. cit., p. 326. A rather special case has been provided by controlled atmosphere storage of apples. Though relatively widely adopted since WW II in the U.S., its use through the mid-1960's was not associated with any particular increase in the proportion of the apple crop stored. Rather, it brought about a significant increase in quality of late season fruit and led to premium prices for storage fruit -- premiums which held up well through the mid-1960's.
(Dana G. Dalrymple, "The Development of an Agricultural Technology: Controlled Atmosphere Storage of Fruit," Technology and Culture, January 1969, p. 47.)

8/ Wilcox and Cochrane, op. cit., p. 326. The situation may be less pressing in the special case where the primary effect of the technology is to reduce costs (which in turn is capitalized into the value of assets). "The individual whose resource inventory has increased in value can transfer the greater post-innovation wealth if he chooses to do so." But where "...the capitalized value of assets has decreased, the resource owner cannot transfer his pre-innovation wealth to other industries." (Heady, op. cit., p. 306.)

9/ Hertford, op. cit., p. 1175; Johnston, op. cit., p. 286; John W. Mellor, "Growth of the Market and the Place of Agricultural Development in Low Income Nations," Cornell International Agricultural Development Mimeograph 22, June 1967, pp. 9-10.

10/ Jack Baranson, "Economic and Social Considerations in Adapting Technologies for Developing Nations," Technology and Culture, Winter 1963, p. 28.

11/ Kendrick, op. cit., p. 1066.

12/ Wilcox and Cochrane, op. cit., p. 327.

13/ Moore, op. cit. (see fn. 28, chp. III), p. 267; John W. Mellor, The Economics of Agricultural Development, Cornell University Press, 1966, p. 157.

14/ Gruen, op. cit. (see fn. 4, chp. II), pp. 838-851.

15/ Ibid.

16/ Philippe P. Leurquin, "Cotton Growing in Colombia: Achievements and Uncertainties," Food Research Institute Studies, 1966 (No. 2), pp. 128, 148.

17/ Ruttan, op. cit. (see fn. 2, chp. II), p. 740.

18/ Also see Marion Clawson, "The Implications of Urbanization for the Village and Rural Sector," in Social Problems of Development and Urbanization (Vol. VII of Science, Technology and Development), Washington, 1963, pp. 49-50.

19/ William McD. Herr, "Technological Change in the Agriculture of the United States and Australia," Journal of Farm Economics, May 1966, pp. 267-268.

20/ R. J. Forbes, The Conquest of Nature: Technology and Its Consequences, Praeger, 1968, pp. 42-43.

21/ Heady, op. cit. (1962; see fn. 3 this chapter), p. 10.

22/ Ibid., p. 184.

23/ Rosenberg, op. cit. (see fn. 1, chp. II), p. 516.

24/ T. W. Schultz, "Production Opportunities in Asian Agriculture: An Economist's Agenda," University of Chicago, Agricultural Economics Paper No. 68:12, revised July 12, 1968, p. 17. (Reprinted in Development and Change in Traditional Agriculture: Focus on South Asia, Michigan State University, Asian Study Center, Occasional Paper, November 1968.)

25/ Claude Moisy, "Enough Wheat for Export?" Ceres, July-August 1968, p. 30.

26/ Dorris D. Brown, "Capital Formation and Agribusiness in India," Columbia Journal of World Business, January-February 1969, p. 62.

27/ Moore, op. cit. (see fn. 28, chp. III), p. 467.

28/ Jones, op. cit. (1965; see fn. 9, chp. II), p. 106.

29/ Cited in I. M. Destler, "Ecological Imbalance -- Man's Pressure on the Land," International Agricultural Development, February 1968, p. 18.

30/ For further discussion, see the following reports based on a Conference on "The Ecological Aspects of International Development Programs": The Unforeseen International Ecologic Boomerang, supplement to Natural History, February 1969, pp. 41-72 (reprints available from the Conservation Foundation, Washington, D.C.); and "Development in the Poor Nations: How to Avoid Fouling the Nest," Science, March 7, 1969, pp. 1046-1048.

31/ Rosenberg, op. cit. (see fn. 1, chp. II), p. 531.

32/ Baranson, op. cit. (1967; see fn. 3, chp. I), p. 524.

33/ H. W. Singer, "The Distribution of Gains Between Investing and Borrowing Countries," American Economic Review, May 1950 (as reprinted in Readings in International Economics, ed. by R. E. Caves and H. G. Johnson, Irwin, 1968, pp. 316, 317).

34/ Ibid.

35/ For further discussion, see: Singer, op cit., pp. 306-317; Robert E. Baldwin, "Export Technology and Development from a Subsistence Level," The Economic Journal, March 1963, pp. 80-92; Jones, op. cit., p. 108."

36/ Ruttan, op. cit. (1960; see fn. 3, chp. II), p. 737.

V. HIGH YIELDING VARIETIES OF GRAIN*

During the late 1960's a great deal of public attention has been given to the role of improved varieties of wheat and rice in expanding food production in the less developed nations. Although some of the public statements have been overdone (e.g. "miracle seeds"), the new varieties have indeed brought about significant changes in agriculture.

A. Background of New Varieties

Breeding for improved varieties of grains has been underway for many years in the developed world, but it is only since WW II that significant progress has been made in more than a handful of the less developed nations.[1]

1. Foundation-Sponsored Research

National breeding efforts received a significant boost with the establishment of the Agricultural Program of the Rockefeller Foundation in Mexico in 1943. Emphasis was placed on developing improved varieties of corn and wheat as well as vegetables. Extension of the Mexican work subsequently led to the establishment of other programs in South America and India, and eventually to an International Wheat Program. The Central American corn project, established in 1954, grew into an International Corn Program. In 1966, the two programs were merged with the formal establishment of the International Maize and Wheat Improvement Center (CIMMYT — Centro Internacional de Mejoramiento de Maiz y Trigo) in Chapingo, Mexico. The Center, now jointly financed by the Rockefeller and Ford Foundations, presently has programs underway throughout the world.[2]

Foundation interests in rice date back to 1952 when Rockefeller sent a preliminary study team to the Far East. It was subsequently decided that Rockefeller and the Ford Foundations would join forces to establish an International Rice Research Institute (IRRI) in the Philippines. The Institute was organized in 1960 and dedicated in 1962. Starting in 1965, scientists employed by the Institute and the Foundations were stationed at key research centers in other countries; they have become intimately involved with the research programs of the host nations.[3]

Thus it can be seen that the two Foundation-sponsored efforts in grain improvement are closely intertwined with national programs.[4] Part of the main concept is that the Institutes do the "basic" research work, while "adaptive" research -- tailoring of the variety to local conditions -- is done at the national level. Not all the new grain varieties have been developed under this cooperative program, but a surprising number tie in in some way -- through parent stocks, training of researchers, etc. Accelerated breeding programs are now underway at many points throughout the less developed world.

---

*References and notes for this chapter are found on pp. 48-51.

2. Spread of New Varieties

There has been a sharp expansion in the area planted to improved varieties of grains in the less developed nations within just the last few years. Most of the increase has been in improved varieties of wheat and rice. Precise data are not available on the exact area planted, but a general picture for wheat and rice is provided in Table 2. Essentially all of the area reported planted to wheat and rice is in Asia (primarily in India and Pakistan), but shipments of improved seeds have been made to many regions.

In India and Pakistan, plantings of improved varieties of corn and other grains expanded as follows: 5/

Crop Year	Corn	Others*
	---- acres ----	
1966/67	563,000	388,000
1967/68	1,012,000	2,519,000
1968/69 (goal)**	(3,000,000)	(5,000,000)

* Grain sorghum, spiked millet, and barley; almost entirely in India.
** Considerable shortfall expected because of dry weather.

In some countries, plantings of the improved varieties represent only a small proportion of the total planted to that crop. In others it is beginning to reach an appreciable proportion: around 20% in the case of wheat in India and Pakistan in 1968. In any case, there has been a striking -- possibly unparalleled — increase in the area planted to new varieties.

B. Nature of Technical Changes

The use of the new varieties entails substantial changes in the agricultural functions involved and the technical base needed.

1. Functions Involved

The main functional changes involved center around production, harvesting, and marketing. Improved varieties of grains require a "package" of inputs if they are to attain their maximum production potential. While any variety will respond to inputs such as fertilizer and water, they will do so only up to a point, and then diminishing returns set in. The new varieties have a much higher response threshold.

This is most easily seen with respect to fertilizer. Unimproved native varieties have only a limited response to fertilizer in terms of yield: much of the added growth goes into the stalk, which, as it becomes taller,

Table 2. ESTIMATED AREA PLANTED TO NEW HIGH-YIELDING VARIETIES OF WHEAT AND RICE IN THE LESS DEVELOPED NATIONS

Crop Year	Acres (rounded)		
	Wheat 1/	Rice 2/	Total
1964/65	negl.	negl.	negl.
1965/66	23,000	14,000	37,000
1966/67	1,554,000	2,343,000	3,897,000
1967/68	9,658,000	6,762,000	16,420,000
1968/69 (goal)	(14,750,000) 3/	(12,300,000)	(27,050,000)

Notes:

1/ Essentially all Mexican or Mexican-type varieties. Excludes Mexico (where an average of 1.85 million acres was planted to improved varieties during the 1960-64 period).

2/ Primarily International Rice Research Institute varieties (IR-8, IR-5), but also includes substantial quantity of: (1) ADT-27 and Taichung (Native) I in India; and (2) BPI-76 in Philippines. Does not include area planted to long-standing improved local varieties in Ceylon and Taiwan.

3/ Plus an unestimated quantity of improved wheat in Nepal.

Source:

Dana G. Dalrymple, "Imports and Plantings of High-Yielding Varieties of Wheat and Rice in the Less Developed Nations," International Agricultural Development Service, December 17, 1968, 18 pp.; as updated for India and Afghanistan. (The reference to wheat acreage in Mexico in fn. 1 was obtained from figures compiled by Reed Hertford of the Economic Research Service.)

is more apt to lodge (bend or break). On the other hand, the high-yielding
varieties have semi-dwarf characteristics: as fertilizer is added, the
response is in terms of grain yield, not stalk growth. Furthermore, the
partly-dwarfed varieties have a stiffer stalk.6/ Thus the point of diminishing returns becomes much higher for fertilizer.

The same is true of water. Consequently, most of the new varieties of
wheat and rice are grown on irrigated land. Recently, however, a new dryland wheat variety has been developed (the wheat still requires water
but can be planted earlier and matures before the summer dry weather sets
in). It, too, requires a package of inputs, including increased fertilization for maximum payoff.7/

Other inputs required include deeper plowing, drill or straight row planting (wheat), improved insect and disease control, more attention to weed-
ing, better management, etc.8/

Even without adequate inputs, the new varieties generally slightly outyield native stock. It is reported that the new dryland wheat, for instance,
will by itself increase yields 15 to 25% over older varieties. But if no
provision is to be made for increased inputs for most of the new varieties,
it may not be worth bothering with them. Thus the adoption of new varieties almost always goes hand in hand with the adoption of new cultural
practices and new inputs.

Use of new production technologies also leads to a need for changes in
harvesting and marketing. The new varieties, especially rice, are quicker
maturing than native varieties. This means an earlier harvesting period;
in the case of rice this can come during the wet season, which in turn
steps up the need for improved drying equipment. Also the increase in
yields can overwhelm existing harvesting and marketing techniques, creating
a need for further changes.

2. Need for Technical Base

The preceding changes in functions must, to be carried out effectively,
be built on a sound technical base. First, the seeds themselves must be
available and of proper quality. High quality seed multiplication requires
exacting technical and ethical standards. These are often difficult to
meet under the conditions that exist in the less developed nations.
Second, manufactured inputs must be available for production purposes.
These include: (1) inputs for direct use such as fertilizer, insecticides
and pesticides; and (2) inputs for the manufacture or application of the
previous items (electricity or gasoline engines may, for instance, be
needed to drive the water pumps for irrigation). Thirdly, manufactured
inputs may be necessary for harvesting, drying, and storage of grains. In
addition, improved communications and transportation may be essential.
Thus, a fairly elaborate technical base -- both in terms of knowledge and
physical inputs -- is necessary to capitalize on the new varieties.

C. Adoption Process for New Varieties

As with other technologies the adoption of new varieties has moved unevenly. Here we shall consider some of the major characteristics of the process.

1. Factors Influencing Rate of Adoption

The early adopters of the new varieties tend, as might be expected, to be better-than-average farmers. They are also usually the larger farmers. In Mexico, for instance, wheat was most quickly adopted by the large, irrigated holdings of the commercially-oriented wheat farmers. This is not, however, always the case: in Kenya in 1967 about two thirds of the 300,000 acres planted to corn was in the lands of small farms, while in Turkey the 420,000 acres planted to wheat were divided up among approximately 60,000 farmers. All told, the new varieties are probably more neutral with respect to farm size than many other technologies.9/

The rate of adoption of new varieties varies widely -- both in terms of type of grain and farm and region involved. With respect to the grains, Mexican progress in wheat has not been matched by progress in corn; Indian progress in wheat has not yet been matched by progress in rice. Similarly, adoption has moved faster in some regions than others.

Barker has identified nine factors which he feels have influenced the rate of adoption of new rice varieties:10/

- water control
- insect and disease problems
- availability of complementary inputs (seed, fertilizer, labor, credit, etc.)
- quality of farm management
- farm institutional structure
- relative advantage of new over existing varieties
- acceptability of the quality of the new grain
- availability of marketing resources
- government institutional structure, pricing policy, and initiative

Water control is needed because the new shorter stemmed varieties cannot survive under flooded conditions: at present about 20% of the rice-growing area in tropical Asia is irrigated. More attention is needed for insect and disease control for any crop grown under intensive high-input conditions; moreover, some of the varieties (particularly IR-8) may not have as much natural resistance as local varieties.

In some regions where the new varieties have been grown successfully for several years, the problems may take on a different complexion. Such is the case in the Thanjavur area of Madras state in India. There the expansion in the use of the new varieties may have hit a plateau because of problems of reviving an ancient irrigation system, tight official control over the rice price (the price in Madras City is reportedly the lowest in any major Indian city), uncertainty over land tenure, and problems in making improvements on rented land.11/ Variations of some of these problems may influence the rate of adoption elsewhere.

2. Role of Government in Adoption

The major role in the spread of the new varieties has been played by the national governments of the LDC's in cooperation with IRRI, CIMMYT, and/or the Agency for International Development (AID).

National political leaders have taken a particular interest in the variety programs:12/

> Afghanistan. Former Prime Minister Maiwandal was so impressed with the production potential of the Mexican wheats and with the urgent need to arrest Afghanistan's growing dependence on imported wheat that he assessed each of the Ministries 2.5% of its current year's development budget to create a fund to launch an accelerated wheat-production program.
>
> India. C. Subramaniam, former Food and Agriculture Minister, took advantage of the food crisis to mobilize support for and launch the accelerated food-production effort responsible for much of India's gains.
>
> Turkey. Prime Minister Demirel feels strongly enough about the crash program in wheat production, initiated at his behest less than two years ago, to have it directed and monitored from his office.

Vigorous support has also been provided by President Marcos in the Philippines and Prime Minister Senanayake in Ceylon.13/

The motivation for such support may not be entirely altruistic. In addition to obvious economic and social advantages to increased production, some political leaders see them as a path to re-election. At least such seems to be the case for President Marcos and Prime Minister Senanayake.14/

Government support has come at many levels -- from exhortations to increased supplies of technical inputs. But because the new varieties require a "package" of inputs, the programs generally tend to be comprehensive in nature. This means accompanying the seeds with: increased quantities of inputs such as fertilizer, broadscale educational efforts, attention to increasing credit supplies, establishment of demonstration plots, initiation of price support and purchase programs, and many other steps.

In some cases these programs were virtually a cooperative effort of the national government and AID. Such was the case for rice in the Philippines, Vietnam, and Laos, and for wheat in Turkey. In the Philippines, among other activities, AID developed a do-it-yourself rice kit which has proved very popular. AID has also been connected with new variety introduction in other nations such as India and Pakistan.15/

Burma is somewhat an exception in that it has largely made its way alone. The route, if not the temper, has been much the same:

> The government has exerted great efforts to induce the Burmese farmer to expand IR-8 production. It has lauded the qualities of IR-8, offered the farmer important material inducements, and mobilized the entire government apparatus and controlled press to support the program.16/

Once initial adoption is secured, continuing government support is necessary to keep the program going. The Philippines has, for example, developed a wide range of activities since 1967.17/ Other nations may find it all too easy to let things slide.

3. Permanency of Adoption

The adoption of new varieties is not a permanent thing. Growers may try a specific variety for a year or so and then decide to drop it with no particular capital loss.

Reasons for dropping a variety probably tend to center around economic matters. Two surveys of farmers in the Philippines who decided not to grow improved rice varieties another season indicated that over 50% of the reasons were due to low price or added expenses; another 10 to 15% related to the added labor involved. A Burmese village reduced acreage because of the poor demand on both the free and black markets.18/

But if growers should drop a variety, there is the possibility that they may continue to use some of the improved practices, with beneficial effects on yields. Moreover, it is quite possible that some of the growers may adopt subsequently improved varieties.

Thus for any one grower, use of new varieties does not represent a locked-in technology, but one where there is, or is likely to be, a continual state of flux.

D. Impact of the New Varieties

The effects of the new variety package of technology are multifold. Some are quite apparent already; others are just beginning to emerge. Here we shall concentrate on the economic and social/political effects at the farm level.

1. Economic Impact

Economic effects of the new varieties center about their effect on production and their consequent influence on financial returns.

a. Agricultural Output

The new varieties have had both qualitative and quantitative impacts on output and have also influenced cropping patterns.

(1) **Quantitative Effect**. The increase in yields accruing from the use of new varieties is quite variable. It is one thing to talk about yields attained under experimental conditions or by top growers, and quite another to talk about average increases actually attained by ordinary growers (who may not fully follow recommendations [19]). On the whole yield increases associated with the new variety package have probably averaged 50 to 100% if the comparison is drawn with conventional cultivation in the better growing areas. In other instances, of course, the increases may be quite different (and they might vary between wheat and rice).[20]

Not all arable land is planted to the new varieties. In 1967/68, new varieties accounted for about (a) 6% of the area planted to rice in South and Southeast Asia and (b) about 16% of the area planted to wheat in South and West Asia.[21] The proportions in individual nations varied widely. Just what percentage will be reached over the next few years will depend on a number of conditions. Since most of the varieties to date (except for the new dryland wheat developed in Lebanon) are grown to best advantage under irrigation, the area of irrigated land available will set an upper limit. The actual figure might well be less than this due to other cultural influences, as well as economic and social factors. On the other hand, the new varieties which have a shorter growing season will increasingly make double cropping possible.

Analyses carried out by the AID missions in India and Pakistan in mid-1968 shed some light on the role played by new variety technology. The studies were cast in terms of the factors contributing to the record increase in production in 1968 (1967/68 for rice) over 1967. Their relative influence was estimated to be as follows:[22]

	India	West Pakistan
	-- percent --	
High-yielding varieties including fertilizer and irrigation	30.3	15.4
Increased fertilization and irrigation of local varieties	28.2	7.5
Expansion of area	6.2	30.0
Weather (other)	35.4	47.1
TOTAL	100	100

It is not clear what portion of the expansion of area, if any, was due to the new varieties.

More generally, it has been estimated that during the 1968/69 crop year improved varieties may have increased rice output about 7% and wheat output 20% in the Asian area, compared to what production would have been without them.23/ As the area planted to new varieties expands they will, of course, play a greater role in increasing production.

(2) _Qualitative Effect_. In terms of demand, the new varieties of grains are often -- if not nearly always -- considered to be of lower quality than local varieties. There are at least three interrelated reasons for this: (a) genetic characteristics of quality may actually be less desirable, and/or (b) the new varieties may require improved methods of harvesting, drying and storing which are not as yet available, and/or (c) the difference may be more imagined than real. We shall look briefly at the first two.

Genetic characteristics of the grain may indeed differ. In the case of rice, texture and taste differences can lead to lower milling and eating qualities -- influencing both domestic and export markets.24/ Some of the Mexican wheats initially used in South Asia varied from local preferences in terms of grain color (they were red rather than white) and baking qualities. Breeding programs, however, are expected to solve many of these problems. Current rice varieties will soon be replaced with new strains developed at IRRI and elsewhere: an IR-8 cross with Basmati, for example, appears promising for export purposes. Numerous white-grained types of wheat have been developed for use in South Asia. More are to come.25/

Problems with harvesting, drying and storing can arise from at least two sources. One is the fact that the sharply increased yields from the new varieties may simply swamp existing harvesting, drying and storage facilities; this in turn may lead to a consequent loss of grain quality. Or, as in the case of rice, there may be a problem of timing:

> The IR-8 was harvested approximately one month earlier
> than local varieties during the end of the wet season.
> Rice sold wet in the field received a price discount
> well in excess of normal drying charges due to the risk
> involved in drying and milling before the rice spoiled.26/

To some extent, the difficulties involved with earlier harvesting of rice could be avoided if better equipment or facilities were available. But all too often they are not. Increased attention is being given to these matters at IRRI and elsewhere.

(3) _Influence on Cropping Patterns_. The preceding quantitative and qualitative effects could have pronounced effect on cropping patterns.

In the short run, the increased returns available from the high-yielding varieties have in some cases led to a shift in production from other crops to rice and wheat. In India, for instance, during the fall of 1968 farmers appeared to show a preference for wheat while the area planted to some other crops such as gram, barley, and pulses showed a small reduction. Alternatively, short-term success with grains may lead nations to overlook other promising crops: one crop economist claims that in Vietnam, for example, preoccupation with rice has led to the neglect of bananas as a potential export crop.27/

Over the longer run, the increased output possible with the new varieties, and the impetus they give to multiple cropping, could well increase grain supplies in some areas to the point where some land is freed or shifted to other crops. This appears to have occurred already among some rice farmers in the Philippines and wheat growers in Mexico (in the latter case, the shift was encouraged by the government's price support program). 28/ The other crops might well take the form of fruits and vegetables, or feedgrains for livestock production. In other words, the new varieties may in effect lay the ground work for the diversification of agricultural production. 29/

   b. Financial Returns

The net returns to farmers from using the new varieties are influenced by changes in returns and costs. It was suggested in Chapter IV that increased returns are most likely to go to those who first adopt a technology, whereas later adopters may find little if any improvement. Our emphasis here will necessarily be on the shorter run.

      (1) Changes in Farm Prices and Costs. The increased output associated with the new variety technology lays the base for increased income in the short run. The big questions concern the direction and magnitude of the price changes. These are directly determined by two factors: the quality of the product and the speed with which production increases.

There is, as we have noted, at least a temporary grain quality differential -- one which has not favored the new varieties. In the Philippines, as of late 1968, IR-8 rice was selling at about 20% below local varieties on the open market, even though the government buys at the same price. Out of 153 farmers surveyed in the Philippines during the wet season in 1967, 148 reported a lower price. In India, the reddish grain produced by the Mexican types of wheat often sells at 10 to 15% less than the best white grain Indian types. The international situation in late 1968 appeared to be particularly difficult for IR-8: one big sale made by the Philippines to India late in October was concluded at a price which reportedly represented a loss. Discounts for the new varieties are not, however, the rule in every market. 30/

The price of grain is, of course, influenced by the quantity available -- both nationally and on the world market -- and the nature of demand. During the late 1960's world production of wheat and rice increased substantially. The demand for these two commoditites, moreover, is generally inelastic. 31/ Thus it is not surprising that there was a decrease in some prices during the latter part of 1968. The FAO world export price index for rice, for example, moved as follows: 32/

Jan.	145	May	149	Sept.	145
Feb.	150	June	148	Oct.	139
March	158	July	149	Nov.	139
April	152	Aug.	147	Dec.	141
				(Jan.	138)

Similarly, the wholesale price index for wheat in India during the June to November period was over 12% below that of a year earlier. 33/ Thus, while providing the basis for an increase in production, the new varieties also likely contributed to a drop in prices.

In some of the less developed nations, the price drop was mitigated at the farm level by government price support and purchase program -- such as that for wheat in India or rice in the Philippines. But this was done at no little strain to government treasuries. The Philippines, in fact, was led to propose an international rice agreement. Production increases in Mexico threatened to create surpluses and in 1966 the government reduced support prices for irrigated corn and wheat grown in leading areas.34/

The added inputs involved in the production of new varieties clearly will raise the costs of production per acre, though not necessarily per ton. In one study in the Philippines, the increase in variable costs per hectare for growing BPI-76 and IR-8 instead of native varieties under traditional practices brought the total costs up by 1/2 to 3/4; costs per cavan of output, however, were little different (and were less for IR-8).35/ Comparable illustrative data are not at hand for wheat.

(2) Changes in Farm Income. It has been widely assumed that the increased returns from growing the new varieties have exceeded the costs. Incomes have probably generally been increased in the short run. Yet there is little solid evidence on this point.

The returns on rice have not been uniform either seasonally or by market. Farm management studies on rice in the Philippines in 1967 suggested that net returns per hectare for IR-8 were two-thirds higher than traditional varieties during the wet season but only slightly higher during the dry season.36/ The situation in Burma varies by market: as of late 1968 it was (a) profitable to raise IR-8 for sale to the government, because the purchase price was the same as for other types of rice,37/ but (b) not profitable to grow it for the free or black markets because of a lack of demand brought on by quality problems.

Returns can also be viewed on a macro or micro basis. The gross value of the increased wheat production in Turkey during the 1967/68 season was estimated at $23.6 million, while total additional costs to farmers were placed at $18.0 million, giving very roughly a net return of more than $5 million. A linear programming study of foodgrain production in the Punjab in India in the mid 1960's, however, suggested that net income per acre would not be significantly increased until fertilizer levels were increased substantially and capital was not a constraint.38/ Much more study is needed on the nature of short and long-run changes in farm income.

2. Social/Political Impact

Social/political problems can arise from the fact that certain groups may not share evenly in the benefits accruing from the new varieties. Within regions there are farmers who may not adopt the new varieties because of economic or other reasons; similarly there may be differences in rate of adoption between regions. While those who adopt the varieties will face, as we have seen, further adjustments, their economic situation may well be better than those who did not adopt them.

Where this is so, there may be a growing or widening economic gap between sectors of the population. This may be less of a problem with the wheat and rice varieties than with more mechanical technologies because they can be relatively widely adopted. In Turkey, for instance, it has been reported that farmers who did not adopt the new wheats were "... those who because of inadequate rating could not borrow money, or those who were rational non-adopters..." 39/

But there is an added dimension to the problem. Not all new varieties have moved as quickly as wheat and rice. Corn in Mexico is a prominent example. Although improved hybrid corn varieties were developed in Mexico along with wheat, a much lower portion of total area has been planted to them. Part of the reason is that corn is the staple crop of the small low-income farmer who usually doesn't have irrigation. Moreover, he may not have the opportunity or resources to buy new seed each year, not to mention the other inputs. The result has been that small farmers have largely been bypassed by technological change. 40/

Technology is, moreover, apt to change much more rapidly than institutions. One of the most severe difficulties in many areas concerns land tenure arrangements:

> Feudal land tenure relationships, which were successfully transformed in postwar Japan and Taiwan by decree, are providing wretchedly durable in southeast and south Asia. Unless legislation keeps pace with economic changes, the agricultural revolution will roll by leaving the rural population dispossessed. 41/

The problem in many of these areas is that the pay-off from innovations is capitalized into land values, resulting in increased rents for tenants.

But the adoption of new varieties has not -- unlike many new technologies -- tended to directly result in the displacement of labor. If anything, it may have led to increased employment in the short run, especially in rice areas. The same may be true over the longer run, particularly as increased grain production provides the basis for diversification involving labor-intensive crops. 42/

If there has not been a sharp quantitative drop in employment, however, there may have been a qualitative shift in wealth. The farmers who have been the first to adopt the new varieties have been made, temporarily at least, financially better off. This increased wealth has not necessarily been passed on to the workers in improved wages. The result has been growing social tension in some areas. In the State of Madras in India, for instance, the uneven returns from the new varieties have led to serious clashes between owners or tenants and agricultural laborers: "Wages have increased unevenly throughout the district, and the landless laborers have been agitating for an even bigger share of the new prosperity." 43/

The problem, as Mellor puts it, is that the new technologies may provide their benefits in proportion to landholdings rather than in proportion to labor inputs.44/ Thus if social/political problems of the sort discussed here are to be avoided, landlords may have to adopt a more enlightened policy on land tenure, while farmers may have to provide more equitable wages for landless laborers. Neither will be easily accomplished.

3. **National and International Implications**

We have discussed the effects of the new variety technology largely in terms of their influence with respect to agriculture. There are, of course, many potential effects at the national and international level.

At the national level, supplies of grains will be increased and prices reduced. Depending on the degree to which marketing and food distribution are improved, the result may be a reduction of undernutrition. To the extent that increases in grain supplies permit increased production of other crops, there may also be a reduction in malnutrition. The improvement in nutrition, together with the possibility of reduced imports and increased exports, will contribute to national economic growth. The results, however, may not be favorable: the temporary increase in consumer income (due to lower prices) and in the income of at least some farmers has caused concern about inflationary pressures in at least one nation.45/

Impacts at the international level may be more complex, particularly with respect to trade. As less developed nations increase grain production, they will first move to economic self-sufficiency and then possibly into an exporting situation. Self-sufficiency will mean a reduction in dependence on foodgrain imports -- especially those of a concessionary nature; this in turn may create some short-run adjustment problems for food grain exporting nations (though over the longer run, with economic development, the market for some agricultural products may improve). But as the less developed countries move into the international grain market for grain, they will face a host of new and sophisticated problems -- involving questions of product quality, comparative disadvantage, trade barriers, etc.; in this market the established exporters may well have the edge. Issues of this nature will doubtless be of increasing concern.46/

References and Notes

1/ Exceptions include Taiwan where the Japanese initiated breeding work early in the century (for details, see: S. C. Hsieh and V. W. Ruttan, "Environmental, Technological and Institutional Factors in the Growth of Rice Production: Philippines, Thailand and Taiwan," Food Research Institute Studies, 1967 (No. 3), pp. 331-333; and Raymond P. Christensen, Taiwan's Agricultural Development: Its Relevance for Developing Countries Today, U.S. Department of Agriculture, Foreign Agricultural Economic Report No. 39, 1968, pp. 36-37).

2/ E. C. Stakman, Richard Bradfield and P. C. Mangelsdorf, Campaigns Against Hunger, Belknap Press of Harvard University Press, 1967, 328 pp.; 1966-67 Report Cimmyt, 93 pp; Cimmyt Report, 1967/68, 99 pp; J. George Harrar and Sterling Wortman, "Expanding Food Production in Hungry Nations; the Promise, the Problems," in Overcoming World Hunger (ed. by Clifford M. Hardin), Prentice Hall, 1969, pp. 89-135.

3/ Stakman, op. cit., pp. 285-299; Randolph Barker, "The Role of the International Rice Research Institute in the Development and Dissemination of New Rice Varieties," International Rice Research Institute, (Los Banos), 1968, 47 pp.

4/ AID has contributed funds to both centers to extend research and training programs. Two additional centers are now being organized: the International Center for Tropical Agriculture (CIAT) in Palmira, Colombia, and the International Institute of Tropical Agriculture (IITA) in Ibadan, Nigeria.

5/ Based on: "India PM," op. cit. (see fn. 10, chp. III), p. D-55; "Pakistan Program Memorandum, FY 1970," US/AID Mission, Ralwapindi, Summer 1968, p. A-53; and Foreign Agricultural Service report IN 9025 from New Delhi, February 1969, p. 4.

6/ A triple-dwarf wheat is now under test in India. It will have even greater resistance to lodging than present varieties. (Cimmyt Report, 1967/68, p. 70; "Indian Rice Crop Seen at Record," Journal of Commerce, December 4, 1968.)

7/ "Dryland Wheat Strain Developed in Beirut," Journal of Commerce, September 13, 1968; "New Wheat to Double Yield," New York Times, September 29, 1968. (Also see Najah, A Dryland Wheat, American University of Beirut, Faculty of Agricultural Sciences, Publication No. 27, January 1967)

8/ Randolph Barker and E. U. Quintana, "Farm Management Studies of Costs and Returns in Rice Production," in The Seminar-Workshop on the Economics of Rice Production (December 1967), International Rice Research Institute, p. 45; David S. H. Liao, "Studies on Adoption of New Rice Varieties," International Rice Research Institute, September 1968, Table 13, figure 2; "Intoduction of Mexican Wheat in Turkey, 1967-68," US/AID, Ankara, July 1968, p. 58.

9/ Barker, op. cit. (1968), p. 28; "Kenya Demonstrates for More Corn," War on Hunger, September 1968, p. 8; "Introduction in Turkey," op. cit., p. 23.

10/ Barker, op. cit. (1968), p. 24.

11/ Adam Clymer, "Madras Rice Progress Imperiled," The Sun (Baltimore), November 19, 1968.

12/ Brown, op. cit. (see fn. 2, chp. I), p. 695.

13/ "Rice -- Miracle, Maybe," The Economist (London), October 19, 1968, pp. 54, 57.

14/ Ibid.; Brown, op. cit., p. 695; Joseph Lelyveld, "Food is Key Issue in Ceylon Politics," New York Times, November 11, 1968.

15/ No comprehensive summary of AID's contributions is presently available. However, a broad-scale review of the new varieties was initiated by AID in late 1968 and should be available in 1969. In the interim, the experience in Turkey has been well reported and may be illustrative: see Ralph N. Gleason in "Turkey's 'Green Revolution' in Wheat-Self-Help in Action," War on Hunger, September 1968, pp. 3-5. Further details are provided in "Introduction," op. cit. (see fn. 8, this chapter) and Department of State Airgrams from Ankara: A-1658, November 29, 1968; A-1706, December 17, 1968; and A-68, February 4, 1969.

16/ Department of State Airgram A-364 from Rangoon, November 23, 1968.

17/ James F. Keefer, "An Afterlook at the Philippine Rice Breakthrough," Foreign Agriculture, March 31, 1969, pp. 4-5.

18/ Barker and Quintana, op. cit., p. 46; Liao, op. cit., Table 12; Department of State Airgram A-910 from Djakarta, November 25, 1968.

19/ A survey in India during the summer of 1967 revealed that farmers participating in the high yielding varieties program were applying only about 50% of the recommended dose of nitrogen and about 70% of the recommended quantity of phosphates ("India PM," op. cit., pp. D-8, D-9).

20/ Less spectacular success has been obtained with rice than wheat in India because rice requires more complex management, especially with respect to water (Ibid., p. D-56).

21/ Donald Chrisler, The World Agricultural Situation, U. S. Department of Agriculture, Foreign Agricultural Economic Report No. 50, February 1969, p. 13.

22/ Based on: "India PM," op. cit., pp. D-8, D-9; "Pakistan PM," op. cit., p. A-5.

23/ Joseph Willett and Donald Chrisler, "The Impact of New Varieties of Grain," U. S. Department of Agriculture, Economic Research Service, December 1968 (draft), pp. 13, 14.

24/ A survey of 153 Philippine rice farmers in 1967 indicated that 144 thought that the eating quality was worse, 5 the same, and 4 better (Liao, op. cit., Table 13). For a graphic reaction to the eating qualities of IR-8 in Vietnam, see Peter Kann, "Miracle in Vietnam; New Rice May be Key to Economic Stability After War Ends in Land," Wall Street Journal, December 18, 1968.

25/ Cimmyt Report, 1967/68, pp. 68-69; Barker, op. cit., p. 30; "Rice -- Miracle, Maybe," op. cit., p. 54; Foreign Agricultural Service report AGR-195 from Ralwalpindi, November 7, 1968.

26/ Barker, op. cit., p. 29.

27/ "Commodity Ring: Indian Wheat Sowings Complete," The Journal of Commerce, December 12, 1968; Kann, op. cit.

28/ Liao, op. cit., p. 12; Foreign Agricultural Service report MX 9008 from Mexico City, February 7, 1969.

29/ For further discussion, see Dana G. Dalrymple, The Diversification of Agricultural Production in Less Developed Nations, U. S. Department of Agriculture, International Agricultural Development Service, August 1968, 56 pp.

30/ Randolph Barker, International Rice Research Institute, personal communication, October 31, 1968; Liao, op. cit., Table 13; Cimmyt Report, 1967/68, pp. 68-69; Romeo M. del Castillo, "RP Prospects Not Too Bright," Manila Times, October 29, 1968; AGR-195 from Ralwalpindi, op. cit.; Foreign Agricultural Service report AGR-76 from Manila, September 30, 1968.

31/ The price elasticity of demand for wheat in India, for instance, has been estimated at -0.55 (John W. Mellor and Ashok K. Dar, "Determinants and Development Implications of Foodgrains Prices in India, 1949-1964," American Journal of Agricultural Economics, November 1968, p. 973).

32/ "Rice Price Intelligence," FAO, February 10, 1969, p. 1. By comparison, the monthly indexes for the 1957 to 1965 period varied no more than 9 points per year and on the average varied only 4. (Computed from data in The World Rice Economy in Figures, 1909-1963, FAO, Commodity Reference Series 3, 1965, pp. 107-108.)

33/ Foreign Agricultural Service report IN 9025 from New Delhi, February 6, 1969.

34/ Foreign Agricultural Service report AGR-92 from Manila, December 11, 1968; John C. Scholl, "Mexico's Grain Problem: A Production Boom That Won't Turn Off," Foreign Agriculture, July 3, 1967, p. 7.

35/ Barker and Quintana, op. cit., p. 34.

36/ Ibid., p. 47.

37/ Government purchase prices are not always the same in other nations.

38/ Gleason, op. cit., pp. 3-5; K. S. Mann, C. V. Moore and S. S. Johl, "Estimates of Potential Effects of New Technology on Agriculture in Punjab, India," *American Journal of Agricultural Economics*, May 1968, pp. 281-290.

39/ "Introduction in Turkey," op. cit., p. 46.

40/ See Henry S. Reuss, *Food for Progress in Latin America*, House of Representatives (Subcommittee Print), February 1967, p. 9. The problems of this group, however, are being given special attention by Cimmyt in the Mexican State of Puebla, initiated in 1967 with a grant from the Rockefeller Foundation (*Cimmyt Report, 1967/68*, p. 13). Also see: Donald K. Freebairn, "The Dichotomy of Prosperity and Poverty in Mexican Agriculture," *Land Economics*, February 1969, pp. 36-39.

41/ "Rice - Miracle, Maybe," op. cit., p. 57.

42/ Farm management studies in the Philippines suggest that the labor cost of growing IR-8 and BPI-76 was about 50% higher than for growing a native variety under traditional conditions. (Barker and Quintana, op. cit., p. 34.)

43/ "Madras is Reaping a Bitter Harvest of Rural Terrorism," *New York Times*, January 15, 1969, p. 12; also see Clymer, op. cit. The "rich" landlords, on the other hand, consider themselves in a cost-price squeeze because of more stable rice prices and higher fertilizer costs ("India Dilemma: Farm Riches - For Few Only," *Chicago Tribune*, February 16, 1969). Also see "The Rich Get Richer," *Science News*, April 5, 1969, pp. 335-336.

44/ John W. Mellor, et al., *Developing Rural India: Plan and Practices*, Cornell University Press, 1968, pp. 359-363.

45/ A rather unusual problem is reported in Vietnam. According to Kann: "The Vietcong are buying seed at black market prices and distributing it in areas they control -- while spreading rumors in other areas that IR-8 causes leprosy and impotence" (Kann, op. cit.).

46/ For a more extended discussion of these issues, see: V. W. Ruttan, J. P. Houck, and E. E. Evenson, "Technical Change and Agricultural Trade: Three Examples (Sugar Cane, Bananas, and Rice)," University of Minnesota, Agricultural Economics Staff Paper P68-4, December 1968, pp. 74-75; Lyle P. Schertz, "World Agriculture in the 1970's," U. S. Department of Agriculture, International Agricultural Development Service, February 1969, 21 pp.; and Clifton R. Wharton, Jr., "The Green Revolution: Cornucopia or Pandora's Box?" *Foreign Affairs*, April 1969, pp. 472-473.

VI. MECHANIZATION OF AGRICULTURE*

Technological improvement in agriculture is synonymous, to varying degrees, with the mechanization of agriculture. The tractor is often the symbol of modernization in less developed nations.

A. Background of Mechanization

Mechanization is a broad term. What does it encompass? Here we shall consider it as the application of power provided by an internal combustion engine to field operations. In other words, it is the use of tractors and associated implements in the production and harvesting of agricultural products. Excluded from this definition are animal drawn equipment or stationary power sources (this is not to say that these items are not to be considered forms of mechanization, only that they are not covered here). The definition, however, intentionally allows for the inclusion of small garden-type tractors.

1. Development and Spread

The tractor and associated equipment are hardly new to the developed world. The first gasoline driven tractors in the United States date back to the turn of the century. Widespread commercial use began to climb rapidly after WW I.1/

In the post-war period, the use of tractors in other nations -- particularly in the relatively developed areas -- began to increase also. Through the twenties, many, if not most, of these tractors were supplied by the United States.2/ But gradually thereafter, U. S. exports played a diminishing role as tractor plants were established overseas, in some cases as subsidiaries of U. S. firms. In recent years the Communist nations have become important suppliers of tractors to the less developed world.3/

The numbers of tractors in use in the major regions of the world during the period from 1930 to 1964 are summarized in Table 3. It will be noted that through the late 1950's, over half of the world's reported tractors were in the United States and Canada. Growth in numbers in Europe was slower than in the United States, but by the mid 1960's the total numbers were about equal. The less developed areas of the world -- Latin America, Africa, and the Near and Far East -- accounted for less than 3% of the total in 1930; the proportion did not change markedly until 1957 when it increased to 6% and 1966 when it reached 8%. The market for tractors in some less developed nations such as India was very strong in 1968.4/

A special characteristic of the post-WW II period has been the sharp expansion in the number of garden tractors (power tillers). Although used in Europe since the late 40's, they have continued to grow in popularity. The most significant increase has taken place in Japan where they are used in rice culture: the average number in use during the 1948-52 period in Japan was 62,000; by the mid-1960's this number increased to 2.5 million.5/ They show considerable promise for other wetland areas of Asia.6/

---

* References and notes for this chapter are found on pp. 63-65.

Table 3. ESTIMATED NUMBER OF TRACTORS IN USE IN MAJOR REGIONS OF THE WORLD 1/

Region	1930	1939	1947	1957	1966
			in thousands		
North America	1,020	1,597	2,890	5,207	5,425
Europe	130	265	512	2,619	5,244
USSR	72	523	400	924	1,660
Oceania 2/	32	53	90	300	406
Latin America	20	35	62	287	512
Africa	10	17	30	182	303
Near East	2	5	10	70	135
Far East 3/	1	3	10	45	127
World 3/	1,287	2,498	4,004	9,634	13,951

Notes:

1/ Includes garden tractors in some countries (but not Japan).

2/ Australia, New Zealand.

3/ Excluding Mainland China.

Sources:

1930-1957. "Progress in Farm Mechanization," Monthly Bulletin of Agricultural Economics and Statistics, FAO, May 1966, p.1.

1966. Production Yearbook, 1967, FAO, 1968, pp. 460-468.

2. Power Available

As the numbers of tractors would suggest, there is a wide range in the amount of tractor power available. This becomes more meaningful when expressed in terms of land under field production (arable land and land under permanent crops). Giles has prepared such estimates for the main regions and/or countries of the world. (Figure 3)

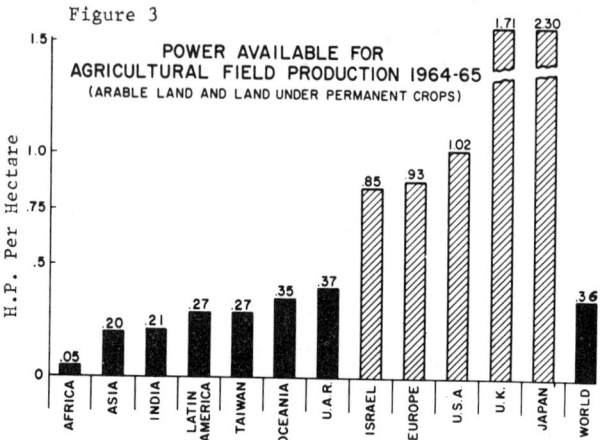

Source: G. W. Giles, "Agricultural Power and Development," in The World Food Problem, The White House, Vol. III, September 1967, p. 172.

He suggests that the minimum amount of power per hectare needed to optimize yields is in the range of 0.5 to 0.8 hp. It will be noted that all the less developed regions or countries on the average fall well below this figure. Within certain areas, of course, the figure may be well above average levels.7/

B. Nature of Technical Changes

The primary characteristic of mechanization is that it represents a substitution of mechanical power for draft animals and in some cases humans. Mechanization can, as Schertz has pointed out: (1) permit the completion of tasks with more precision; (2) accomplish work more quickly; (3) develop resources not presently being utilized; and (4) accomplish tasks not possible with traditional techniques.8/

1. Functions Involved

The functions involved in mechanization run the gamut from field clearing and preparation to road hauling. Actual use depends in part on the size of tractor. In the Allahabad area in India, it has been reported that small riding and walking tractors were used in three main ways: field work, powering irrigation pumps, and carting work (irrigation occupied the greatest number of hours -- more than 40% for many of the tractors). In the Punjab area, larger tractors were primarily used for presowing operations, and to a lesser extent for harvesting. In some regions, road hauling is probably important.9/

Comprehensive information is not available on the types of implements used with tractors. In Allahabad, intensive use is made of only a few: a tiller, a pump, and a trailer. More implements are expected to move into general use, but only over a period of time.10/ Giles has prepared a general priority list of farm operations which may give some idea of relative importance:11/

Priority	Operations and Equipment
	Seed and Plant Bed Preparation
1	Plows, mouldboard or disc (principally for dry land)
1	Power tiller (primarily for wet land)
3	Peg harrow
	Seeding and Fertilizer
1-2	Seed drill or row planter with fertilizer distributor
3	Broadcast fertilizer distributor
	Pest Control
1	Knapsack power duster-sprayer
2	Row cultivator
3	Tractor mounted duster-sprayer
	Harvesting and Threshing
1	Reaper and stationary thresher
2	Self-propelled combines

2. Need for Technical Base

One of the basic prerequisites for mechanization is the existence or development of a technological base. Men must be trained in the proper use and repair of machinery. Spare parts and shop facilities and tools must be available. All of this seems very obvious and is easy to take for granted, but it has been a major stumbling block for mechanization programs in country after country.

As Azam explains the repair situation alone in West Pakistan:

> ...even for the minor repairs the supplying firm has to be contacted, because only a few diesel technicians are available, who are employed by the

tractor firms. The time taken by the agents of
the firm to reach the farm and undertake repairs
causes delay in performing the operations, especially during the cultivation or sowing season, which
ultimately causes financial loss. Most of the dealers
do not import sufficient spare parts along with
tractors, which creates bottlenecks later on.12/

Indeed, if there is any one characteristic that appears virtually histori-
cally consistent for the less developed nations, it is the sight of
tractors standing idle for lack of mechanics and/or some key part. In
1930, Hindus spoke of fleets of disabled tractors dotting the Russian
landscape. And it has recently been reported that in India more than 40%
of the imported tractors are idle due to lack of spare parts.13/

C. Adoption Process for Machinery

How is the adoption of machinery related to size of operation? Why is
machinery purchased? What is the acquisition process?

1. Size and Nature of Farms Involved

Whether mechanization is adopted -- and the form it takes -- is in general related to the size of farms involved.

In India, a study of mechanization in the Punjab area in the early 1960's
indicated that tractor owners were both the larger and wealthier farmers.
For the nation as a whole, it has been suggested that:

- 11% of the area is in small farms (less than
  5 acres) which need good hand implements.

- 25% is in medium size farms (5 - 15 acres) which
  need bullock implements.

- 25% is in medium large farms which could use
  power tillers and matching equipment.

- 40% is in large farms which could use tractors.

Similarly, in the case of West Pakistan -- where a relatively extensive
agriculture is practiced -- Giles indicates that holdings of under 25
acres can be classed for animal-powered farming, and those above 25 acres
for tractor-powered farms.14/

A somewhat similar situation is found in Latin America. In Colombia,
Lidman noted that tractor distributors would sell on credit to owners of
as few as 37 acres (15 ha.). The tractor dealers assume that the smaller
owners will be able to do rental or custom work; the State Bank is less
optimistic and insists that a farmer be able to employ his equipment full-
time on his own holding. Adams observed in Colombia, however, that small
size of farm operation was more a barrier to ownership than to mechanization because of the availability of rental equipment.15/

In Africa, de Wilde indicates that hand cultivation will continue to be
necessary or economic where:

> (1) topography makes plowing too difficult or un-
> desirable; (2) demographic pressures make holdings
> too small for cultivation with tractors or for sup-
> porting draft animals; (3) the forest cover is so
> dense as to preclude economic clearing except for
> hand cultivation; (4) tree crops can be grown ...;
> (5) enough labor can be hired at reasonable cost to
> cope with peak labor requirements.16/

From all of this it is not surprising that the first to mechanize --
topographical and biological factors permitting -- are the large farmers,
followed more slowly by those who are smaller. The pace would depend on
the availability of rental equipment and the suitability of relatively
low-cost garden tractors. There is, however, a minimum size below which
it may pay to stay with livestock and hand power.

Those in the forefront of tractor purchases are not exclusively full-time
farmers. The group may include businessmen and political leaders who
have farm holdings. This appears to be the case in Laos and West Pakistan,
and is probably true in other nations.17/

2. Reasons for Adopting

The main reasons for adopting mechanization are economic, although social,
political and other reasons are sometimes important. Some of the main
types of benefits which may be presumed to exist by adopters are as
follows:

- Increased Production. Mechanization can conceptually
make it possible to (a) increase the amount of land farmed as well as
(b) to increase the intensity of land use. With mechanization the farmer
may farm more land than he can handle with livestock: a tractor can plow
far more land in a day than can an ox. Furthermore, the farmer can add
to his land resources by plowing land that couldn't be tilled by oxen.
He can also add to his resources by reducing the amount of land necessary
for forage production for the draft animals. Mechanization also makes it
possible to increase output on existing lands. With tractor-drawn equipment, farms can plow deeper and carry out other tillage jobs better. Work
can be done much faster. Timing of planting and cultivation can be improved.

- Reduced Production Costs. It is usually expected that
mechanization will lower per unit costs by increasing yields and reducing
labor costs. Although labor is a relatively abundant good in most less
developed nations (Japan being an exception), there are generally peak
periods of demand -- at planting and harvest time -- when it is apt to be inadequate for the tasks at hand. Moreover, during this period wages are
apt to increase. Therefore, many growers hope to use machinery to break
labor bottlenecks. Reduction of labor may also lead to management savings.

- Reduced Harvesting Losses. Mechanization permits more rapid farm operations. This may be of considerable importance in harvesting as well as planting. For most crops, harvesting is best carried out during a certain limited period; where it is difficult to do this because of labor shortages or unfavorable weather conditions, losses can result. In West Pakistan during the 1968 spring wheat crop harvest, sufficient and qualified labor was not forthcoming on larger farms, resulting in a long drawn out threshing period and an increased threat of spoilage; this has increased interest in combine harvesters.[18]

- Reduced Drudgery. Mechanization, especially where it replaces hand labor, can result in a significant reduction in drudgery. This can have economic implications that might not be immediately considered: for instance, in Northeastern Japan it is estimated that farmers become unable to operate animal-driven plows, heavy loaded carts or weeders in the field by the time they are 44; mechanization can, in effect, expand their useful lifespan.[19] Still, as deWilde has noted:

> Mechanization is unlikely to be economic unless the farmer understands from the beginning that it is not designed to lighten his total work burden but rather to enable him to work more and produce more.[20]

Farmers and government officials may also be attracted to mechanization for prestige or psychological reasons. Mechanization is a tangible, physical example of modernization. There have also been less innocent reasons for mechanization: the tractor, for example, played a key role in the collectivization of Soviet agriculture.[21]

3. Acquisition Process

Mechanization involving large equipment can require a fairly high initial investment. Full utilization is normally necessary to make the investment worthwhile in a less developed nation.

Thus, large tractors are normally purchased outright by the larger farmers or by those who have worked out arrangements to do contract work for other farmers. While much of the contracting work is probably done on a relatively informal basis for neighboring farms, more formal contracting services are provided in some nations by private firms or the government. A recent FAO report indicates that private contracting services have been developed successfully in a number of countries, including Argentina, Ceylon, Chile, Kenya, Malaysia, the Sudan and Thailand. The same report, however, states that nearly all of the government contracting services have failed to cover costs and have had to be abandoned or continuously subsidized.[22]

Other procedures for providing multifarm use include pooling and/or cooperative ownership. Where the land holdings are too small and scattered to permit economical operation, they may be consolidated. Structurally different forms of farming such as settlement schemes, collective, or state farms also permit mechanization, though they may be undesirable on other counts.

Thus the advantages of mechanization can be made available to a wider range of farmers than those who are able to buy the equipment. Just how much of this is done in the less developed nations, however, is not known. A report from Laos indicates that more farmers there would use custom services if they were reasonably sure of the availability of service and could secure credit to defray costs until harvest. Similarly, a survey of rice farmers in the central Philippines revealed that over 50% would use a garden tractor for land preparation if they could conveniently hire it done.23/

D. Impact of Mechanization

How well do the presumed reasons for the adoption of mechanization match up with reality? What are economic and social/political effects? How do they balance each other?

1. Economic Impact

The economic effects of mechanization may be primary and secondary in nature. For our purposes, the primary effects will be considered as those that accrue at the farm level and have their main effect on income. The secondary effects are those which have broader interrelationships.

a. Primary Effects of Mechanization

There seems to be relatively little quantitative information available on the economic effects of mechanization in the less developed nations. This is not entirely surprising in view of the relatively low level of mechanization in the past and the wide variety of conditions in these countries. Several studies have recently been conducted in Asia which, when published, may shed more light on the matter. We shall divide our few comments here between production and harvesting.

(1) Production and Costs. Mechanization is presumed to lead to increased output and lower costs. There is particularly little documentation on the former point. Typical is the rather general statement by Ballis that: "The impression and estimates of the farmers are unanimous that tractor cultivated land produced significantly more..."24/

There is more information available on costs of machinery relative to livestock. At a conference in Japan in December 1967, a number of studies pertaining to Asia were summarized. The results varied widely, depending on the type of equipment, its purchase and maintenance costs, and the season. In general, however, garden tractors seemed to be relatively more efficient during the wet season, while larger tractors fared better during the dry season when their greater power is needed for tillage.25/

In some instances, the bullock has been the cheapest source of power in Asia, but its position has been eroded by the more exacting requirements of an increasingly intensive agriculture, and increased labor and feed costs. One Philippine study indicated that one of the factors that prompted rice growers to give up carabaos for garden tractors was the high cost and short average useful life of the carabao (many died of chemical poisoning).26/

In some areas, such as India, the cost of owning a tractor has been high largely because they were used for only limited functions (tillage and transport) a limited portion of the year. Increasing the number of operations performed over a greater portion of the year will make them more economical.27/

(2) <u>Reduced Losses</u>. Mechanization makes it possible to carry out harvesting operations more quickly and with less loss. Nervik and Haghjoo indicate, for instance, that combining can reduce the harvest period from 2 to 3 months to 1 to 2 weeks, greatly reducing the risk of losses through spoilage and weather hazards. Giles suggests that in West Pakistan, mechanical harvesting and threshing can save the estimated 10 to 15% of the grains now being lost.28/

Proper timing is particularly important with rice:

> Where early harvesting is carried out, wastage will be high because of the presence of chalky and immature grains. Where harvest is late the greater number of sun-cracked grains results in a high percentage of brokens and wastage in the milled product.29/

Work underway at IRRI, under AID sponsorship, has produced a promising drum-type field thresher which will, for the first time, handle wet rice.30/

(3) <u>Farm Organization</u>. Mechanization can both stimulate an increase in farm size or lead to increased intensity through multiple cropping. Leurquin reports that in Colombia increased use of machinery in rice production has tended to expand the optimum size of farms. On the other hand, deWilde suggests that in Africa, the introduction of machinery sooner or later makes it necessary to intensify production.31/

In any case, it is clear that a primary effect of mechanization is to give the farmer more time -- which he may spend on enlarging and/or intensifying his farm operation. At the same time mechanization is likely to bring the farmer further into the market economy: as Moerman observed in a Thai village, "Shortly after the introduction of the tractor ...price and profit became major standards for crop selection."32/

b. Secondary Effects of Mechanization

With production and harvesting changes, along with market entry, there will be a greater need for improvements in marketing. This may be due to the interrelated nature of technical improvements. A recent example from the American midwest is illustrative:

> Because of changes in the method of harvesting, from ear corn to shelled corn, this once storable commodity is now a perishable commodity requiring

almost instant conditioning to preserve it for
year-around consumption.33/

In the case of the LDC's, we have seen that crop changes -- such as increased production of IR-8 -- will lead to a need for improved drying and storing facilities.

Alternatively, the need for marketing improvements may come from a different direction. It may be that if society is to benefit from the fruits of production technology, it will be necessary to make public investment in marketing infrastructure such as roads, communications, etc.

Beyond marketing, there are many other potential secondary economic effects of national concern. The influence on farming patterns can work several ways. On one hand, for example, the decrease in livestock numbers brought about by mechanization means a decrease in the direct supply of manure, meat and hides. On the other hand, production of other types of livestock may be increased: a study of mechanization in the Punjab and Bahawalpur areas of India revealed that:

> ...with the eviction of bullocks from farms, increased quantities of fodder and husk became available, which resulted in an increased number of milch /milk/ cattle on farms.34/

Or interest may be stimulated in poultry or hog farming as elsewhere in Asia.35/

Numerous other economic impacts could be traced out -- such as the usual need to import petroleum products and parts, bringing in foreign exchange considerations -- but perhaps enough have been mentioned to suggest the complex and intertwined nature of secondary influences.

2. Social/Political Impact

Most of the really disruptive problems of mechanization show up at the social/political level. These may take many forms but we shall concentrate on two of the most critical: labor displacement and by-passed groups.36/

Perhaps the biggest difficulty with mechanization is in the displacement of labor. This does not always occur -- as in the case where mechanization leads to intensified production -- but is usually at least a strong potential problem. From a qualitative point of view, there will be a lessening of demand for unskilled labor and an increase in demand for scarce skilled labor -- machinery operators and repairmen. Quantitatively, there still may well be an overall decrease in demand for labor. Bose and Clark, in a study of mechanization in the Punjab area of West Pakistan, learned that the labor force had been reduced about 50% from the pre-mechanization period.37/

These changes can lead to alterations in the structure of rural society. Farmers dislocated from the farm may first move to the local village,

and from there to urban centers.[38] Or the move may be directly from
farm to city. In any case the result will be the same for the urban area:
a further influx of relatively untrained individuals and a dearth of jobs.

Not all of the effects may be quite so obviously disruptive as the labor
question. It may be that social problems arise because mechanization has
little to contribute to improving the status of the more disadvantaged
farmers. As Lidman put it:

> Unfortunately, these machines will not and cannot
> be instrumental in the solution of Peru's most
> pressing agricultural problem: the social and
> economic state of the subsistence minifundia.[39]

Similarly, tenants may find that a disturbingly large portion of any profits they might make from mechanical improvements are passed on to land
owners through a rise in land rentals.[40] The fact that these farmers
remain static while others push ahead can widen the economic and social
gap between the upper and lower sectors of society.

3. Balancing Economic and Social Considerations

In total, we see that mechanization may bring pronounced economic benefits
in the form of increased output at lower costs. This can mean more food
at lower cost to society. On the other hand, it may bring a displacement
of rural labor. How do we balance economic benefits against social costs?

One of the first studies to treat this matter, at least in part, was conducted by Lidman in two areas of Peru. Mechanization in two valleys was
evaluated in terms of economic rationality for farmers and social welfare.
He concluded that in one region, the Mantaro Valley, the use of tractors
was both economically and socially desirable. However, in another region,
the Jequetepeque Valley, the use of a combine was economically rational
for farmers, but of "...doubtful value to society as a whole."[41]

Economic and social aspects of mechanization in West Pakistan have recently
been examined in two excellent -- though as yet unpublished -- papers: one
by Bose and Clark and the other by Johnston and Cownie.[42] Both studies
independently indicated that the private benefit to those who introduce
tractors is much greater than the social benefit. Private returns were enhanced by the fact that tractors were being imported duty free at official
exchange rates. Social benefit was reduced, of course, because of the
adverse effect on employment. Both studies went on to suggest that a maximum rate of expansion might not be optimal -- that a slower pace of mechanization may be called for.

Ruttan adds that the problem is not so much one of balance as it is one
of distorted economics. Mechanization may be encouraged in situations
which otherwise might not be economic as a result of subsidized interest
rates, unrealistic exchange rates, or other institutional devices.[43]

These are important questions and deserve further study.

References and Notes

1/ For a brief history, see E. M. Dieffenbach and R. B. Gray, "The Development of the Tractor," in *Power to Produce* (The Yearbook of Agriculture), 1960, pp. 25-45.

2/ In 1926, the United States provided the following proportions of tractors purchased: France 80%, Australia 80%, Mexico 100% (Robert E. Linneman, "A Case for Minimum Marketing Efforts in International Markets," *Mississippi Valley Journal of Business and Economics*, Fall 1965, pp. 75, 79, 83). Virtually all of the tractors used in the Soviet Union during the mid-1920's came from the U. S. (Dana G. Dalrymple, "American Technology and Soviet Agricultural Development, 1924-1933," *Agricultural History*, July 1966, p. 193).

3/ India for instance, plans to make the following purchases from Communist nations in 1969: 6,500 from the Soviet Union (6,000 of 18 hp. size), 5,000 from Czechoslovakia, 3,000 from East Germany, and 500 from Rumania (Department of State Airgram A-5 from New Delhi, January 2, 1969).

4/ According to the U. S. AID Mission in India, there was a backlog of 20,000 to 30,000 orders for tractors as of the summer of 1968 ("India PM," *op. cit.* (see fn. 8, chp. III), p. D-61). For a recent review of the tractor supply situation, see /John Parker/ "Tractor Takeover in South Asia," *The Farm Index*, January 1969, pp. 15-17.

5/ *Production Yearbook, 1967*, Food and Agriculture Organization, 1968, p. 470.

6/ A large body of information on the use of tillers and tractors in Asia is presented in *Expert Group Meeting on Agricultural Mechanization* (December 1967), Asian Productivity Organization (Tokyo), Vol. I, June 1968, Vol. II, October 1968.

7/ G. W. Giles, "Agricultural Power and Development," in *The World Food Problem*, The White House, Vol. III, September 1967, p. 177, 179.

8/ Lyle P. Schertz, "The Role of Farm Mechanization in the Developing Countries," *Foreign Agriculture*, November 25, 1968, p. 3. (Also published as "Food, Man and Machines," *War on Hunger*, January 1969.)

9/ John S. Ballis, "Summary of Tractor Evaluation Project, Allahabad Agricultural Institute," Progress Report No. 14, March 1967, pp. 8-9; Bhagat Singh, "Economics of Tractor Cultivation - A Case Study," *Indian Journal of Agricultural Economics*, January-March 1968, p. 85.

10/ Ballis, *op. cit.*, p. 12.

11/ Giles, *op. cit.*, pp. 189-190.

12/ K. M. Azam, "Economics of Farm Mechanization," in his *Planning and Economic Growth*, Maktaba-tul- Arafat (Lahore), 1968, p. 117.

13/ Maurice Hindus, <u>Red Bread</u>, Jonathan Cape, 1931, pp. 357-358 (I have discussed Soviet maintenance problems elsewhere: see "The American Tractor Comes to Soviet Agriculture: The Transfer of a Technology," <u>Technology and Culture</u>, Spring 1964, pp. 206-207); <u>Expert Group</u>, <u>op. cit</u>.: Vol. I, p. 307, Vol. II, p. 116.

14/ A. C. Pandya, "Rationale for Agricultural Implements and Power Programmes for Five Year Plans," <u>Farm Journal</u> (India), November 1967, p. 15; Singh, <u>op. cit</u>., p. 84; Giles, <u>op. cit</u>., p. 2.

15/ Russell M. Lidman, "The Tractor Factor: Agricultural Mechanization in Peru," <u>Public and International Affairs</u> (Princeton University), 1968 (No. 1) p. 18; comments of Dale W. Adams, Office of Program and Policy Coordination, Agency for International Development, January 1969.

16/ John C. DeWilde, <u>Experiences with Agricultural Development in Tropical Africa</u>, Johns Hopkins University Press, 1967, Vol. I, pp. 96-97.

17/ Department of State Airgram A-1010 from Vientiane, April 8, 1968; S. R. Bose and E. H. Clark, "Some Basic Considerations On Agricultural Mechanization in West Pakistan," Williams College, November 1968 (unpublished manuscript), p. 50.

18/ "Economic Trends and Their Implications for the United States," American Embassy, Rawalpindi, July 20, 1968.

19/ Expert Group, <u>op. cit</u>., Vol. II, p. 6.

20/ de Wilde, <u>op. cit</u>., p. 116.

21/ Dalrymple: <u>op. cit</u>. (1964), pp. 191-214; <u>op. cit</u>. (1966), pp. 187-206.

22/ "Raising Agricultural Productivity," <u>op. cit</u>. (see fn. 1, chp. I), p. 96. For an account of the sociological relationships between farmers and tractor owners in a Thai village, see Michael Moerman, <u>Agricultural Change and Peasant Choice in a Thai Village</u>, University of California Press, 1968, pp. 69-79. For an example of a program that failed, see Herman J. Van Wersch, "Rural Development in Morocco: Opération Labour," <u>Economic Development and Cultural Change</u>, October 1968, pp. 33-49.

23/ Airgram A-1010 from Vientiane, <u>op. cit</u>.; S. S. Johnson, E. U. Quintana, and Loyd Johnson, "Mechanization of Rice Production," in <u>The Seminar-Workshop on the Economics of Rice Production</u> (December 1967), International Rice Research Institute, pp. 3-18.

24/ Ballis, <u>op. cit</u>., p. 8.

25/ Expert Group, <u>op. cit</u>., Vol. II, Chp. IV, pp. 85-112.

26/ Tagumpay-Castillo, <u>op. cit</u>. (see fn. 20, chp. III), p. 291.

27/ /Howard E. Ray/ New Opportunities Through IADP for Growth in India's Agriculture, Report to the Secretary of Agriculture by the Ford Foundation in India, 1968, pp. 16, 56-58.

28/ Ottar Nervik and E. Haghjoo, "Mechanization in Underdeveloped Countries," Journal of Farm Economics, August 1961, pp. 663-664; G. W. Giles, "Towards a More Powerful Agriculture," distributed by The Planning Cell, Agriculture Department, Government of West Pakistan, Lahore, November 1967, p. 12.

29/ "Rice in the World Food Economy," The State of Food and Agriculture, 1966, Food and Agriculture Organization, 1966, p. 167.

30/ See "Thresher Developed in the Philippines," War on Hunger, January 1969, p. 15.

31/ Philippe P. Leurquin, "Rice in Colombia: A Case Study in Agricultural Development," Food Research Institute Studies, 1967 (No. 2), p. 268; de Wilde, op. cit., p. 103.

32/ Moerman, op. cit., p. 68.

33/ James Hanscom, "Corn Exports to Double by '72," Journal of Commerce, September 9, 1968.

34/ Azam, op. cit., p. 123.

35/ Expert Group, op. cit., Vol. II, p. 7.

36/ For a discussion of other social/anthropological influences of mechanization in a Thai village, see Moerman, op. cit., pp. 62-87.

37/ Bose and Clark, op. cit., p. 35.

38/ This, at least, has been the response to the early stages of mechanization in the Mississippi Delta (see Richard H. Day, "The Economics of Technological Change and the Demise of the Sharecropper," American Economic Review, June 1967, pp. 441-442).

39/ Lidman, op. cit., p. 30.

40/ Leurquin, op. cit. (1967), p. 268.

41/ Lidman, op. cit., pp. 15, 29.

42/ Bose and Clark, op. cit., 53 pp., esp. p. 52; Bruce F. Johnston and John Cownie, "The Seed-Fertilizer Revolution and the Labor Force Absorption Problem," Food Research Institute, Stanford University, January 1969 (unpublished manuscript) 38 pp.; and letter from Johnston, February 11, 1969.

43/ Letter from Vernon W. Ruttan, University of Minnesota, March 18, 1969.

VII. POLICY IMPLICATIONS OF TECHNOLOGICAL CHANGE*

The preceding chapters have suggested a number of questions of public policy concerning technological change. In this chapter we shall approach the matter in four stages: need for evaluation, criteria for evaluation, policy problems, and concluding remarks.

A. Need for Evaluation

Technological change in agriculture, while resulting in great benefits to society, also can bring with it substantial problems. Many of these have already been noted. Here we shall summarize and comment on some of those which are most important from a policy viewpoint.

1. Benefits of Technology

There is little question of the overall benefits to society from technological change. Quantitatively, more food and agricultural products are available at lower cost. Qualitatively, the food may be of the type more desired by consumers and may be of higher nutritional value. In addition, relatively new foods may appear or traditional products may become available over a longer season. Furthermore, a greater output may be produced with the same or fewer resources, and growers are drawn into the market economy. The combined result of these changes is that economic development is in general stimulated.

2. Problems of Technological Change

The difficulties with technical change in agriculture arise because of the uneven distribution of benefits and the disruption associated with change.

a. Uneven Distribution of Benefits

The distribution of benefits from technological change is uneven (1) within agriculture and (2) between agriculture and the rest of society.

We have seen that within agriculture, the first few who adopt a technology are apt to benefit financially, but as the technology moves into widespread use, this advantage is competed away. Thus those that follow may not receive any positive payoff, but they may be better off than those who were unable to adjust at all.

The benefits of widespread technological advance, over the longer run, flow instead to consumers. Whether they are evenly distributed among all consumers is difficult to say -- but certainly more so than among producers. To a varying extent producers are also consumers of marketed food so some farmers may gain in this way. In any case, the transfer of income is clearly not an orderly process from a welfare point of view:

---

* References and notes for this chapter are found on pp. 74-75.

"...consumers who realize gains in the form of more food for less money are not generally poor and farmers whose income may diminish are not universally wealthy."1/

In assessing the payoff from technological change, a distinction should be drawn between absolute and relative benefits. In absolute terms, nearly everyone is economically as well or better off with technological improvements (the prominent exceptions being entrepreneurs who were unable to adjust and displaced laborers). But it is in the nature of technological change that the benefits accrue relatively unevenly. Thus, as Tweeten and Tyner put it, if farm income has suffered, it is in relative rather than in absolute terms.2/

b. Disruptions Associated with Change

The essence of technological improvement is change: change directed toward economic growth. This process can only, as Heady describes it, "...result in continuous instability of subsistence or primary product industries."3/ The dislocations can manifest themselves in social and political forms.

The relationship between economic growth and political stability is a complex one. Huntington suggests that it may vary with the level of economic development. He points out that at one extreme some measure of economic growth is necessary to make instability possible: in Eric Hoffer's words, there is a "conservatism of the destitute." On the other hand, in countries which have reached a relatively high level of development, a high rate of growth is compatible with political stability.4/

In an underdeveloped society which has started on the path of economic growth, then, problems of political instability may be expected. This is especially true where growth is rapid:

> ...rapid economic growth, far from being the source of domestic tranquility it is sometimes supposed to be, is rather a disrupting and destabilizing force that leads to political instability. This does not mean that rapid economic growth is undesirable. It means, rather, that no one should promote the first without bracing to meet the second.5/

Economic growth and/or political change can also lead to social disruptions. Traditional social arrangements or relationships will be altered and new ones substituted. The adoption of new agricultural technologies almost inevitably means, for instance, that traditional subsistence farmers are drawn into the market economy. But the differing allocation of returns, even in relative terms, can stimulate social unrest.

The various disruptions associated with economic growth may be widespread, but they are perhaps most concentrated in the agricultural sector. Moore, after a massive historical study, has concluded that the poor inevitably bear the heaviest costs of modernization -- whether they occur under socialist or capitalist auspices:

> The only justification for imposing the costs is
> that they would become steadily worse off with-
> out it. As the situation stands, the dilemma is
> indeed a cruel one.6/

All of this may seem a bit extreme if one is thinking of some relatively modest technical changes in agriculture. And it may well be. But
it is not beyond the realm of possibility. We have only to remember
that the use of new varieties of rice has already led to political dis-
turbances in some areas of India. Nearly every change will introduce
some disruptions; attempts should be made to evaluate them in advance
and then, where called for, "brace" to meet them.

B. Criteria for Evaluation

Evaluation of the effects of technological advance necessitates the establishment or recognition of criteria. The criteria, in turn, will depend
on the goals and values of society.

The direction given farm technical advance, as Heady has pointed out,
might differ depending on the specific ends to be achieved. He identi-
fies three potential economic goals:

1. Increasing the total net income of the agricultural
   industry.

2. Increasing the total utility or welfare of individuals
   now in the agricultural industry.

3. Maximizing aggregate economic progress.

These goals, he indicates, are not identical and are not without conflict.7/
Moreover, they suggest a certain tendency to judge growth solely in terms
of economic criteria.

Recently, it has become increasingly recognized that a broader interpretation is needed. The tendency to view development as essentially an
economic process has masked the need to also measure social progress and
social justice in the less developed nations (the latter, of course, would
involve the development of new criteria and measurements). In the fall
of 1968, for example, the Pakistan Planning Commission asked whether the
time had not come to shift the emphasis in the country's economic planning
from growth to social justice; subsequent political events in Pakistan have
underscored this need. A high rate of economic growth, as the National
Planning Association recently pointed out, cannot be sustained for very
long without related political and sociocultural changes. Thus, technological advance needs to be evaluated in terms of its composite economic
social and political effects.8/

It should not be forgotten, however, that despite the many political, social,
and cultural interrelationships, development remains primarily an economic
process. In this context, as Drucker has indicated, "...the essence of

development is not to make the poor wealthy, it is to make the poor productive."9/

It is clearly impossible to say what or how broad a perspective should be taken for judging agricultural technologies; this will depend on the specific circumstances. But it is clear that it might well be desirable to move beyond a narrow input-output analysis to a broader spectrum -- both in the short and long run.

C. Policy Issues

Among the many policy issues that may be raised by the introduction of a new technology, two of the more important from an agricultural point of view are: (1) the choice of factor combinations in agriculture, and (2) distribution of the gains and losses.

1. Factor Combinations and Investments

Evaluation of factor choices can vary depending on whether market conditions or developmental criteria are used.

> For example, firms choose optimal production techniques based upon factor prices (which are usually determined by market conditions), but these do not always reflect the relative scarcity of foreign exchange and capital equipment (which must be taken into account by those concerned with the overall economic development of the country).10/

Even if one starts with development criteria in mind, it must be remembered that farmers will make their decisions on the basis of market conditions.

One of the traditional questions concerns the balance to be struck between investment in biological and mechanical factors. In addition to being influenced by the foreign exchange position of the country -- for mechanization usually involves imports -- this question is influenced to a large extent by the degree of development.

It seems to be generally recognized that (barring unusual circumstances) at low levels of development the nod should be given to biological innovations. This is because (a) they are needed to increase the response of plants to technical inputs or change yield potentials to provide scope for the introduction of engineering improvements, and (b) they often do not require large capital investments or severe displacement of labor.11/

But as agricultural development takes place, and crop changes set the stage for mechanization -- e.g. as increased double cropping made possible by new varieties increases the need for mechanization -- the labor capital cost ratio begins to swing toward mechanization. Mechanization, moreover, can initially be a relatively simple process -- involving the use of power tiller or garden tractor. Then, as conditions call for it, larger tractors can be adopted. At the later stages of development (and in some cases

well before), improvements in crop production may provide the basis for
livestock production; concurrent growth in income provides the basis for
an expansion in demand.

There are, of course, numerous exceptions to a sequence placing crops
first, followed by mechanization and livestock improvement. Irrigation
may be involved at a number of levels and can even be a necessary precedent for crop improvement. Some mechanization may be necessary to clear
land for crops. In any case, the link between crops and mechanization
is much closer than the relationship between mechanization and livestock
improvement.

Emphasis in agricultural development programs has traditionally and rightly been placed on production. But as programs to improve production begin
to pay off, other second stage problems begin to become more important.
Perhaps the most significant from a technological point of view is the
need to improve marketing facilities for both inputs and final products.
Included will be such items as improved roads, storage and processing.
At some point, therefore, a decision will have to be made on the balance
of resources allocated to production and marketing.

In the short run, conflicts can also arise between allocations for consumption and for investment. Should policy-makers, for instance, use scarce
foreign exchange to import food as opposed to fertilizer or (to go a step
further) for machinery for a fertilizer plant? Or at the national level,
should policy makers discourage farm investment in consumption goods such
as household items and encourage investment in farm improvements such as
tubewells?12/

The answer would have to be: some balance. But this may not be easily
arrived at. It is tempting, for instance, to place a relatively low
priority on consumption goods, but they can play an important role in
providing incentives which lead to mobilization of under-employed resources
for increased production and income.13/ The problem is not so much with
long-term goals, which are to maximize welfare, but how to get there.

2. Distribution of Gains and Losses

The distribution of the gains and losses from technical progress provides
one of the main bases for the policy problems of American agriculture.14/
Much the same problem is true in less developed nations which have experienced technological breakthroughs.

a. Distribution Within Agriculture

A policy planning dilemma that faces many less developed nations is whether
to allocate limited resources to (1) the more favored areas or the more
progressive producers, or to (2) spread measures to promote technological advance more thinly over the country. The first measure will generally result in greater economic efficiency but the second has undeniable
social and welfare advantages. The choice is not an easy one.

Conflicting opinions have recently been expressed by two authors which illustrate some of the dimensions of the problems. Etienne, writing in an Indian context, states:

> Regional differences in pace do not matter, provided the nationwide growth rate overtakes and keeps ahead of the population rate. This strategy might be challenged on the grounds that it is antisocial, since it benefits mainly the more favored regions and the middle classes of farm society. Can we approve such an agricultural policy? Yes, for this is the only way out, even from the social standpoint.15/

Schickele, on the other hand, states:

> It surely is not enough to depend only on the top 10% of present producers .... They alone will not be able to sustain economic growth because there will not be necessary concomitant expansion of demand ..../More-over/ whatever head start might be gained by an initial dependence on a small producer elite is fraught with social and political dangers and economic setbacks.16/

One effort which in some ways avoids these two extremes has been conducted in India since 1960. It is known as the Intensive Agricultural Districts Program and involves providing growers with a "package" of improved practices. The program avoids individual selectivity in that it is available to all groups in a chosen area; it avoids a mass approach in that the areas selected were generally among the more productive in the country. Although we do not have a review of the social aspects of the national effort, it is reported that in the Tanjore District by 1965 there had been a broadscale adoption of new practices: both large and small farmer, and owner and tenant participated.17/ The subsequent economic benefits, however, are a moot point: one recent study questions whether the overall program stimulated production increases beyond those which would have occurred anyway.18/ (Part of the problem was that several critical aspects of the original program were not carried out; also, in retrospect, not enough attention was given to the development of new technology.19/) But in view of experience to date, it should be possible to design an improved program.

Still, it may be that programs which stress improved income distribution within agriculture will usually do so at some cost in terms of potential agricultural output. Where this in the case, the challenge will be to strike a balance between the two which will be acceptable in both the short and long run. Further, the policy should encourage adequate technological change in agriculture, yet not lead to a major transfer of income to large or wealthy farmers.20/ All of this, however, is much more easily said than done.

b. Distribution Between Producers and Consumers

There is little than can be done to prevent the flow of benefits from technical change from farm to consumers. But it might be possible, as Ruttan has put it, to design programs which both encourage technological change in agriculture and at the same time slow the transfer of savings from farmers to consumers sufficiently to ease the downward pressure on farm income.[21]/

Tweeten and Tyner have suggested three possible rates of advance that farmers might consider optimum:

1. A zero rate of technological advance, with rising prices and income through a fixed or static supply and inelastic product demand.

2. Maintain farm resource returns at opportunity cost levels without aggregate farm resource adjustments.

3. Allow farm output needs to be produced with 1% fewer conventional inputs per year. The necessary rate of productivity increase is 2.8% annually.

The latter rate most nearly illustrates the route society has adopted.[22]/

The idea of adjusting rates of technological change has been given relatively little thought in the United States and undoubtedly even less in underdeveloped countries. My own reaction is that this route is not a very promising one because (1) it is not necessarily to the direct advantage of consumers as a whole, and (2) it would be very difficult to carry out, in part because of the many sources of technical change.[23]/ Still, some adjustments might be possible, and the matter is important enough to be worth further study.

***

We have discussed only a few of the possible policy issues resulting from technological change. Others — such as renewed interest in rural taxation or overloading of administrative structures -- may well become important in the future. Hopefully the background information provided in this report will be of use in dealing with these and other questions.

D. Concluding Remarks

In a hungry world, the natural inclination is to view technological changes in agriculture largely in terms of their immediate effect on agricultural production and food supplies. But changes which have short-run effect also have longer-term implications. And changes which initially affect food supply set in motion a chain of interrelated economic, social and political factors. Many of these in turn will react back on agriculture.

Emphasis on short-run physical increases in production is a logical first step for nations faced with the threat of hunger. Increasing agricultural production is no small task: in most cases it cannot effectively be carried out if diluted by a myriad of other goals.[24/] But where these production efforts begin to pay off, or in nations with a more advanced agriculture, a broader outlook is called for: social and political factors become of increased importance.

Perhaps both past and prologue for agriculture in the less developed nations are well illustrated by analogy. A recent review of developments in civil engineering in the United States concluded that:[25/]

> /The works/ ...discussed in this book have been triumphs of the human spirit because of the obstacles men overcame in building them; tomorrow's structures must be triumphs for the human spirit, built to enhance life for man in all respects.
>
> These will be built, not by engineers building alone as masters of their specialty but by synthesizers who take into consideration the whole range of human needs and technical possibilities.

To be sure, this situation may seem remote for agriculture in many less developed nations. But it is inevitable that technical change -- long a way of life in the developed nations -- will play an ever-increasing role in the agriculture of less developed countries.

Since technical change inevitably brings disruptions with it, the outlook for the developing nations is hardly one of serenity. But with development, the problems change in character: they change from ones of scarcity to those of abundance. This represents absolute improvement but does not do away with problems of relative imbalance.

The answer is not to try to do away with technical change, or even to try to sharply slow it, but to anticipate the problems that it may bring and plan to alleviate or mitigate them. In carrying out this assessment, it is necessary to look beyond immediate economic impact at the farm level to a longer-term analysis reflecting social and political factors at a regional, national, and possibly international level.

References and Notes

1/ T. W. Schultz, "A Policy to Redistribute Losses from Economic Progress," Journal of Farm Economics, August 1961, p. 557; Heady, op. cit.(1949; see fn. 6, chp. II), p. 311 (source of quote).

2/ L. G. Tweeten and F. H. Tyner, "Toward an Optimum Rate of Technological Change," Journal of Farm Economics, December 1964, p. 1080.

3/ Heady, op. cit. (1949), p. 310.

4/ Huntington, op. cit. (fn. 15, chp. III), pp. 52-53. Quote from Eric Hoffer, The True Believer, New American Library, 1951, p. 17.

5/ Mancur Olson, Jr., "Rapid Economic Growth as a Destabilizing Force," The Journal of Economic History, December 1963, p. 552.

6/ Moore, op. cit. (see fn. 28, chp. III), p. 410.

7/ Heady, op. cit. (1949) pp. 306-307.

8/ Robert E. Asher, "A Development Assistance Program for the 1970's," Brookings Institution, September 1968 (unpublished manuscript), p. 51; Joseph Lelyveld, "Pakistani Economic Panel Questions Own Plans," The New York Times, December 1, 1968; A New Conception of U. S. Foreign Aid, National Planning Association, Special Report No. 64, March 1969, p. 5.

9/ Peter Drucker, "A Warning to the Rich White World," Harper's, December 1968, p. 75.

10/ Baranson, op. cit. (see fn. 3, chp. I), p. 520.

11/ Vernon W. Ruttan, "Engineering and Agricultural Development," University of Minnesota, Department of Agricultural Economics, July 1967, p. 40; Earl O. Heady, "Processes and Priorities in Agricultural Development," in Economic Development of Tropical Agriculture (ed. by W. W. McPherson), University of Florida Press, 1968, pp. 66-69.

12/ Mellor, op. cit. (1966; see fn. 13, chp. IV), p. 124.

13/ Ibid., p. 126.

14/ Heady, op. cit. (1962, see fn. 3, chp. IV), p. 10.

15/ Gilbert Etienne, "Manchala and Pilkhi -- Techniques are Not Enough," Ceres, July-August 1968, p. 43.

16/ Ranier Schickele, Agrarian Revolution and Economic Progress, Praeger (Special Studies Series), 1968, pp. 47-48.

17/ Carl C. Malone, "Some Responses of Rice Farmers to the Package Program in Tanjore District, India," *Journal of Farm Economics*, May 1965, pp. 257, 264, 267. Landless laborers did not participate in this process, resulting in the social problems noted earlier in this report.

18/ Dorris D. Brown, "Agricultural Development in India's Districts: The Intensive Agricultural Districts Programme," unpublished manuscript, 1969, pp. 35-71 (to be published by the Development Advisory Service, Harvard University). Also see Carl Malone, "The Intensive Agricultural District Programme," paper presented at the International Seminar on Change in Agriculture, Reading, England, September 1968.

19/ Critical aspects of the original package plan which were not carried out included: the use of the whole farm plan, large amounts of medium-term credit to extend minor irrigation and improve land, technical engineering assistance, and incentive level foodgrain prices. Consequently, much of the remaining emphasis was placed on increasing the number of extension workers (D. Brown, *op. cit.*, pp. 18-19, 25-32, 115-118).

20/ John H. Sanders and Vernon W. Ruttan, "Another Look at the World Food Problem," *Minnesota Agricultural Economist*, February 1969, p. 3.

21/ Ruttan, *op. cit*. (1960, see fn. 3, chp. II), p. 751.

22/ Tweeten and Tyner, *op. cit.*, pp. 1079-1080.

23/ I have discussed some of the limitations inherent in the Tweeten-Tyner paper, and in an earlier paper by Heady, in several notes in the *Journal of Farm Economics*: "The Public Investment in Agricultural Research and Education: Some comments," November 1965, pp. 1020-1022; "Comments on Public Purpose in Agricultural Research and Education," May 1962, pp. 444-453.

24/ One problem with some development programs is that more emphasis has been placed on the social and political concerns of the national elites than on raising food production (W. David Hopper, "Investment in Agriculture: The Essentials for Payoff," in *Strategy For the Conquest of Hunger* /Proceedings of a Symposium, April 1968/, The Rockefeller Foundation, pp. 104-105).

25/ David Jacobs and Anthony E. Neville, *Bridges, Canals and Tunnels*, Van Nostrand, 1968, p. 128.

VIII. SELECTED BIBLIOGRAPHY*

A. Books

John C. de Wilde, et al., Experiences with Agricultural Development in Tropical Africa, Johns Hopkins University Press, 1967, Vol. I.

R. J. Forbes, The Conquest of Nature: Technology and its Consequences, Praeger, 1968.

Earl O. Heady, Agricultural Policy Under Economic Development, Iowa State University Press, 1962.

Samuel P. Huntington, Political Order in Changing Societies, Yale University Press, 1968.

Herbert F. Lionberger, Adoption of New Ideas and Practices, Iowa State University Press, 1960.

Edwin Mansfield, The Economics of Technological Change, Norton, 1968 (not cited).

John W. Mellor, The Economics of Agricultural Development, Cornell University Press, 1966.

John W. Mellor, et al., Developing Rural India: Plan and Practices, Cornell University Press, 1968.

Ezra J. Mishan, The Costs of Economic Growth, Praeger, 1967.

Michael Moerman, Agricultural Change and Peasant Choice in a Thai Village, University of California Press, 1968.

Barrington Moore, Jr., Social Origins of Dictatorship and Democracy: Lord and Peasant in the Making of the Modern World, Beacon Press, 1966.

Everett M. Rogers, Diffusion of Innovations, The Free Press, 1962.

Prodipto Roy, et al., Agricultural Innovation Among Indian Farmers, National Institute of Community Development (Hyderabad), 1968.

Walter W. Wilcox and Willard W. Cochrane, Economics of American Agriculture, Prentice Hall, 1960.

E. C. Stakman, Richard Bradfield and P. C. Managelsdorf, Campaigns Against Hunger, Belknap Press of Harvard University Press, 1967.

---

*Selected from published references only (except for items listed in Section D.)

B. Bulletins and Reports

Dana G. Dalrymple, *The Diversification of Agricultural Production in Less Developed Nations*, U.S. Department of Agriculture, International Agricultural Development Service, August 1968.

**

Development and Change in Traditional Agriculture: Focus on South Asia, Michigan State University, Asian Studies Center, Occasional Paper, November 1968 (symposium papers).

Expert Group Meeting on Agricultural Mechanization (December 1967), Asian Productivity Organization (Toyko), Vol. I June 1968, Vol. II October 1968.

Five Years of Research on Dwarf Wheats, Indian Agricultural Research Institute (New Delhi), 1968 (not cited).

The Rockefeller Program in the Agricultural Sciences (Progress Report: Toward the Conquest of Hunger), 1965-1966; *Strategy For the Conquest of Hunger* (Proceedings of a Symposium, 1968), The Rockefeller Foundation.

1966/67 Report, Cimmyt, *Cimmyt Report, 1967/68*, International Maize and Wheat Improvement Center (Mexico City).

C. Articles

1. In Books

K. M. Azam, "Economics of Farm Mechanization," in his *Planning and Economic Growth*, Maktaba-tul-Arafat (Lahore), 1968 (originally published in *Trade and Industry*, Karachi, July 1965).

Jack Baranson, "The Challenge of Underdevelopment," in *Technology in Western Civilization* (ed. by M. Kransberg and C. W. Pursell, Jr.), Oxford University Press, 1967, Vol. II.

Walter P. Falcon and Carl H. Gotsch, "Lessons in Agricultural Development - Pakistan," in *Development Policy - Theory and Practice* (ed. by G. F. Papanek), Harvard University Press, 1968.

J. George Harrar and Sterling Wortman, "Expanding Food Production in Hungry Nations; The Promise, the Problems," in *Overcoming World Hunger* (ed. by Clifford M. Hardin), Prentice Hall, 1969.

Earl O. Heady, "Processes and Priorities in Agricultural Development" in *Economic Development of Tropical Agriculture* (ed. by W. W. McPherson), University of Florida Press, 1968.

Nathan Rosenberg, "The Economic Consequences of Technological Change, 1830-1880," in *Technology in Western Civilization* (ed. by M. Kransberg and C. W. Pursell, Jr.), Oxford University Press, 1967, Vol. I.

2. Journal Articles

Jack Baranson, "Economic and Social Considerations in Adapting Technologies for Developing Nations," Technology and Culture, Winter 1963.

Santi Priya Bose, "Characteristics of Farmers who Adopt Agricultural Practices in Indian Villages," Rural Sociology, June 1961.

Dorris D. Brown, "Capital Formation and Agribusiness in India," Columbia Journal of World Business, January - February 1969.

Lester R. Brown, "The Agricultural Revolution in Asia," Foreign Affairs, July 1968.

Dana G. Dalrymple, "The American Tractor Comes to Soviet Agriculture: The Transfer of a Technology," Technology and Culture, Spring 1964.

Dana G. Dalrymple, "American Technology and Soviet Agricultural Development, 1924-1933," Agricultural History, July 1966.

Richard H. Day, "The Economics of Technological Change and the Demise of the Sharecropper," American Economic Review, June 1967.

Zvi Griliches, "Hybrid Corn: An Exploration in the Economics of Technological Change," Econometrica, October 1957.

Zvi Griliches, "Congruence Versus Profitability: A False Dichotomy," Rural Sociology, September 1960.

F. H. Gruen, "Agriculture and Technical Change," Journal of Farm Economics, November 1961.

Earl O. Heady, "Basic Economic and Welfare Aspects of Farm Technological Advance," Journal of Farm Economics, May 1949.

William McD. Herr, "Technological Change in the Agriculture of the United States and Australia," Journal of Farm Economics, May 1966.

Samuel Pao-San Ho, "Agricultural Transformation Under Colonialism: The Case of Taiwan," The Journal of Economic History, September 1968.

Bruce F. Johnston, "Agriculture and Economic Development: The Relevance of the Japanese Experience," Food Research Institute Studies, 1966 (No. 3).

William O. Jones, "Environment, Technical Knowledge, and Economic Development in Tropical Africa," Food Research Institute Studies, 1965 (No. 2).

John W. Kendrick, "The Gains and Losses from Technological Change," Journal of Farm Economics, December 1964.

Martin Kriesberg, "Marketing Food in Developing Nations -- Second Phase of War on Hunger," Journal of Marketing, October 1968.

Journal Articles cont'd

Russell M. Lidman, "The Tractor Factor: Agricultural Mechanization in Peru," Public and International Affairs (Princeton University), 1968 (No. 1).

David E. Lindstrom, "Diffusion of Agricultural and Home Economics Practices in a Japanese Rural Community," Rural Sociology, June 1958.

Philippe P. Leurquin, "Cotton Growing in Colombia: Achievements and Uncertainties," Food Research Institute Studies, 1966 (No. 2).

Philippe P. Leurquin, "Rice in Colombia: A Case Study in Agricultural Development," Food Research Institute Studies, 1967 (No. 2).

Carl C. Malone, "Some Responses of Rice Farmers to the Package Program in Tanjore District, India," Journal of Farm Economics, May 1965.

Ghulam Mohammad, "Private Tubewell Development and Cropping Patterns in West Pakistan," The Pakistan Development Review, Spring 1965.

Ottar Nervik and E. Haghjoo, "Mechanization in Underdeveloped Countries," Journal of Farm Economics, August 1961.

Mancur Olson, Jr., "Rapid Economic Growth as a Destabilizing Force," The Journal of Economic History, December 1963.

Vernon W. Ruttan, "Research on the Economics of Technological Change in American Agriculture," Journal of Farm Economics, November 1960.

Theodore W. Schultz, "A Policy to Redistribute Losses from Economic Progress," Journal of Farm Economics, August 1961.

H. W. Singer, "The Distribution of Gains Between Investing and Borrowing Countries," American Economic Review, May 1950 (Reprinted in Readings in International Economics (ed. by R. E. Gaves and H. G. Johnson), Irwin, 1968).

Daniel W. Sturt, "Producer Response to Technological Change in West Pakistan," Journal of Farm Economics, August 1965.

Luther G. Tweeten and Fred H. Tyner, "Toward an Optimum Rate of Technological Change," Journal of Farm Economics, December 1964.

William and Helga Woodruff, "Economic Growth: Myth or Reality," Technology and Culture, Fall 1966.

3. Magazine Articles

James M. Blume, "Consequences of the Green Revolution," War on Hunger, August 1968 (not cited).

Peter F. Drucker, "A Warning to the Rich White World," Harper's, December 1968.

Magazine Articles cont'd

Gilbert Etienne, "Manchala and Pilkhi -- Techniques are Not Enough," Ceres (FAO Review), July-August 1968.

Theodore R. Freeman, Jr., "New Wheats Power Bigger Pakistan Harvests," Foreign Agriculture, April 7, 1969 (not cited).

Ralph W. Gleason, "Turkey's 'Green Revolution' in Wheat: Self Help in Action," War on Hunger, September 1968.

James F. Keefer, "An Afterlook at the Philippine Rice Breakthrough," Foreign Agriculture, March 31, 1969.

Claude Moisy, "Enough Wheat for Export?" Ceres (FAO Review), July-August 1968.

Lyle P. Schertz, "The Role of Farm Mechanization in the Developing Countries," Foreign Agriculture, November 25, 1968.

John C. School, "Mexico's Grain Problem: A Production Boom That Won't Turn Off," Foreign Agriculture, July 3, 1967.

Clifton R. Wharton, Jr., "The Green Revolution: Cornucopia or Pandora's Box?" Foreign Affairs, April 1969.

**

"Rice - Miracle, Maybe," The Economist, October 19, 1968.

4. Newspaper Articles

Adam Clymer, "Madras Rice Progress Imperiled," The Sun, Baltimore, November 19, 1968.

Peter R. Kann, "Miracle in Vietnam; New Rice May be Key to Economic Stability After War Ends in Land," Wall Street Journal, December 18, 1968.

Joseph Lelyveld, "Food is Key Issue in Ceylon Politics," New York Times, November 11, 1968.

Joseph Lelyveld, "Pakistani Economic Panel Questions Own Plans," New York Times, December 1, 1968.

**

"Madras is Reaping a Bitter Harvest of Rural Terrorism," New York Times, January 15, 1969.

5. Articles in Other Publications

Randolph Barker and E. U. Quintana, "Farm Management Studies of Costs and Returns in Rice Production," in The Seminar-Workshop on the Economics of Rice Production (December 1967, International Rice Research Institute.

Articles in Other Publications cont'd

Marion Clawson, "The Implications of Urbanization for the Village and Rural Sector," in Social Problems of Development and Urbanization (Vol. VII of Science, Technology and Development), Washington, 1963.

G. W. Giles, "Agricultural Power and Equipment," in The World Food Problem, The White House, Vol. III, September 1967.

S. S. Johnson, E. U. Quintana, and Loyd Johnson, "Mechanization of Rice Production," in The Seminar-Workshop on the Economics of Rice Production (December 1967), International Rice Research Institute.

**

"Raising Agricultural Productivity in Developing Countries Through Technological Improvement," in The State of Food and Agriculture 1968, FAO, September 1968.

D. Unpublished (Mimeographed) Papers*

Randolph Barker, "The Role of the International Rice Research Institute in the Development and Dissemination of New Rice Varieties," International Rice Research Institute, September 1968.

Swadesh R. Bose and Edwin H. Clark II, "Some Basic Considerations on Agricultural Mechanization in West Pakistan," Williams College, November 1968.

Dana G. Dalrymple, "Imports and Plantings of High-Yielding Varieties of Wheat and Rice in Less Developed Nations," U. S. Department of Agriculture, International Agricultural Development Service, December 1968.

Bruce F. Johnston and John Cownie, "The Seed-Fertilizer Revolution and the Labor Force Absorption Problem," Stanford University, Food Research Institute, January 1969.

David S. H. Liao, "Studies on Adoption of New Rice Varieties," International Rice Research Institute, November 1968.

V. W. Ruttan, J. P. Houck and R. E. Evenson, "Technological Change and Agricultural Trade: Three Examples (Sugarcane, Bananas, and Rice)," University of Minnesota, Agricultural Economics Staff Paper P68-4, December 1968.

Lyle P. Schertz, "World Agriculture in the 1970's," U. S. Department of Agriculture. International Agricultural Development Service, February 1969.

---

* Most of these papers were prepared for seminars or talks and will eventually be published.

D. Unpublished (Mimeographed) Papers cont'd

Clifton R. Wharton, Jr., "Risk, Uncertainty, and the Subsistence Farmer: Technological Innovation and Resistance to Change in the Context of Survival," Agricultural Development Council, December 1968.

Joseph Willett and Donald Chrisler, "The Impact of New Varieties of Grain," U. S. Department of Agriculture, Economic Research Service, December 1968.

World Food Supply

An Arno Press Collection

Agricultural Production Team. **Report on India's Food Crisis & Steps to Meet It.** 1959

Agricultural Tribunal of Investigation. **Final Report.** Presented to Parliament by Command of His Majesty. 1924

Bennett, M. K. **The World's Food:** A Study of the Interrelations of World Populations, National Diets and Food Potentials. 1954

Bhattacharjee, J. P., editor. **Studies in Indian Agricultural Economics.** 1958

Brown, Lester R. **Increasing World Food Output:** Problems and Prospects. 1965

Brown, Lester R. **Man, Land & Food:** Looking Ahead at World Food Needs. 1963

Christensen, Raymond P. **Efficient Use of Food Resources in the United States.** Revised Edition. 1948

Crookes, William. **The Wheat Problem.** Revised Edition. 1900

Developments in American Farming. 1976

Dodd, George. **The Food of London.** 1856

Economics and Sociology Department, Iowa State College. **Wartime Farm and Food Policy,** Pamphlets 1-11. 1943/44/45

Edwards, Everett E., compiler and editor. **Jefferson and Agriculture:** A Sourcebook. 1943

Famine in India. 1976

Gray, L. C., et al. **Farm Ownership and Tenancy.** 1924

Hardin, Charles M. **Freedom in Agricultural Education.** 1955

High-Yielding Varieties of Grain. 1976

[India] Famine Inquiry Commission. **Report on Bengal.** 1945

Johnson, D. Gale. **Forward Prices for Agriculture.** With a New Introduction. 1947

King, Clyde L., editor. **The World's Food.** 1917

Marston, R[obert] B[right]. **War, Famine and our Food Supply.** 1897

Mosher, Arthur T. **Technical Co-operation in Latin-American Agriculture.** 1957

The Organization of Trade in Food Products: Three Early Food and Agriculture Organization Proposals. 1976

Projections of United States Agricultural Production and Demand. 1976

Rastyannikov, V. G. **Food For Developing Countries in Asia and North Africa:** A Socio-Economic Approach. Translated by George S. Watts. 1969

Reid, Margaret G. **Food For People.** 1943

Schultz, Theodore W., editor. **Food For the World.** 1945

Schultz, Theodore W. **Transforming Traditional Agriculture.** 1964

Three World Surveys by the Food and Agriculture Organization of the United Nations. 1976

U. S. Department of Agriculture, Agricultural Adjustment Administration. **Agricultural Adjustment:** A Report of Administration of the Agricultural Adjustment Act, May 1933 To February 1934. 1934

U. S. Department of Agriculture. **Yearbook of Agriculture, 1939:** Food and Life; Part 1: Human Nutrition. 1939

U. S. Department of Agriculture. **Yearbook of Agriculture, 1940:** Farmers in A Changing World. 1940

[U. S.] House of Representatives, Committee on Agriculture. **Oleomargarine.** 1949

[U. S.] National Resources Board. **Report of the Land Planning Committee. Part II.** 1934